SHUTTLE, HOUSTON

MY LIFE IN THE CENTER SEAT OF MISSION CONTROL

—— **PAUL DYE** ——

hachette
BOOKS

New York

Hachette Books
Hachette Book Group
1290 Avenue of the Americas
New York, NY 10104
HachetteBooks.com
Twitter.com/HachetteBooks
Instagram.com/HachetteBooks

First Edition: July 2020

Hachette Books is a division of Hachette Book Group, Inc.
The Hachette Books name and logo are trademarks of Hachette Book Group, Inc.

The publisher is not responsible for websites (or their content) that are not owned by the publisher.

The Hachette Speakers Bureau provides a wide range of authors for speaking events. To find out more, go to www.hachettespeakersbureau.com or call (866) 376-6591.

Print book interior design by Six Red Marbles, Inc.

Library of Congress Cataloging-in-Publication Data
Names: Dye, Paul (Paul F.), author.
Title: Shuttle, Houston: my life in the center seat of Mission Control / Paul Dye.
Description: First edition. | New York: Hachette Books, 2020. | Includes index.
Identifiers: LCCN 2019052492 | ISBN 9780316454575 (hardcover) | ISBN 9780316454544 (ebook)
Subjects: LCSH: Dye, Paul (Paul F.) | Lyndon B. Johnson Space Center—Officials and employees—Biography. | United States. National Aeronautics and Space Administration—Officials and employees—Biography. | Lyndon B. Johnson Space Center. Flight Operations Directorate—History. | United States. National Aeronautics and Space Administration—History. | Space Shuttle Program (U.S.)—History. | Aerospace engineers—United States—Biography. | Astronautics—United States—History.
Classification: LCC TL789.85.D88 A3 2020 | DDC 629.4092 [B]—dc23
LC record available at https://lccn.loc.gov/2019052492

ISBNs: 978-0-316-45457-5 (hardcover), 978-0-316-45454-4 (ebook)

Printed in the United States of America

LSC-C

10 9 8 7 6 5 4 3 2 1

Dedication...

It's more than just a word for anyone involved in human spaceflight operations—it is a way of life. No matter what you did to make Shuttle operations a success, whether you were a flight controller, a design engineer, an astronaut, a trainer, or a flight planner—you dedicated your life to the project. The job required total commitment for many years, and countless hours of hard work.

Because of your contributions, this book is dedicated to you. Only some of your names appear in these pages, but that's mostly because in thirty years, tens of thousands of men and women served in the many positions required to make the Space Shuttle fly. So while your name might not appear explicitly, if you served as an EECOM, a FIDO, a GC, a DPS, or a CAPCOM—or any of the hundreds of jobs that were essential to each and every flight—you are in this book. If you were an instructor or a flight planner—you are in this book. If you worked in the MER, at the Cape, or in a POCC—you are in this book. I was proud to work with every one of you, and I'm excited to be able to share some of our stories with a larger audience.

And to the hundreds of thousands of engineers and technicians throughout the country (and the world) who made the vehicles and got them ready to fly—I add my humble thanks to those of the rest of the world. You have served humankind in the most amazing profession... exploration!

And to my family and friends who have waited patiently for this book to appear—thank you for giving me the time it took to finally tell these stories.

—Iron Flight

Contents

SHUTTLE, HOUSTON

Introduction

It was almost forty years ago that I first stood in the doorway of NASA's Mission Operations Control Room—the heart of the building known as MCC, or Mission Control Center—and felt a chill run through my body. This place, this room—history had been made here. In this room, Christopher Kraft and Gene Kranz led teams that conducted the Gemini missions. These missions taught America how to spend weeks in space, how to walk in space, and how to rendezvous with another spacecraft in orbit. These missions transformed our space experience from lobbing men into orbit to one of actually flying into space.

But Gemini was just a rehearsal for the big show—the Apollo trips to the moon. I watched those missions in my youth, riveted to the black-and-white TV set at home. I watched as Apollo 8 orbited the moon on Christmas Eve. I watched as Neil Armstrong stepped off the ladder of Apollo 11 and onto the lunar surface in the summer of the next year. And I watched as the men and women in this very room worked round the clock to bring home the crew of Apollo 13 after their spacecraft was crippled on the way to the moon.

That room went on to be the place where engineers and scientists used to explore the moon on later Apollo missions—then went on to link up with the Soviets, and to build America's first space station, Skylab. That room had filled my childhood, and now here I was in 1980, standing on the threshold of this almost sacred space. The carpet and textured walls made it a quiet place—and the fact that everyone inside was talking softly on a headset made it even quieter. Many have compared it to stepping into a cathedral, and that feeling was palpable; this was a place where great things could—and did—happen. It demanded respect, and maybe even a bit of reverence. The lighting was low to make it easier to read the digital displays and the arrays of event lights. Flight controllers reached quietly and calmly for books from behind their consoles—stacking them in piles of open binders to cross-check information. There was no panic—it was all matter of fact. But it was exciting beyond measure—for I knew that they were training to fly the very first Space Shuttle mission.

I remember going with one of my mentors to visit the Control Center—he needed to talk to a flight controller who was working a simulation for the first Shuttle flight. The controller was in a back room somewhere, doing a support job, and it was common to bother someone in that position during training. We walked into the back room—a room full of consoles and an even larger array of books than were in the front room—and found the person we needed to see. He looked up and waved us silent—he was obviously listening to the voice loops. "What can I do for you?" he asked after a minute or so. "I've got about five minutes before we come up on Hawaii, and when we get the signal, I'll have three minutes to verify the platform alignment before we lose them, and they will be doing the deorbit burn after that. So, we've got to get that right—but I have a minute—so what do you need?" I realized then and there that multitasking was going to be an important skill to perfect if I wanted to be successful.

The Space Shuttle was the culmination of all we had learned about operating in space with human beings—and it was a huge leap in capability and technology. Yes, the Saturn V that sent men to the moon for the Apollo missions was taller, heavier, and could accelerate the spacecraft to much higher velocity. And yes, that spacecraft went far higher than low-Earth orbit; it went to the moon. But as marvelous as it was, the Apollo spacecraft elements were still simple compared to the mechanical, electronic, and aerodynamic complexity of the Space Shuttle. The Space Shuttle was capable of carrying up to 50,000 pounds into orbit, in a payload bay almost 60 feet long and 12 feet in diameter. Not only could it loft that large cargo, it could bring a huge amount back. Up until the Shuttle, returning items from space was almost as costly as lofting them into orbit in the first place—and little had ever come back, save a few hundred pounds of moon rocks—and, of course, the astronauts.

Not only could the Shuttle take a lot to orbit and back, the spacecraft itself was reusable, unlike all the Apollo, Gemini, and Mercury capsules that now decorate museums across the country. A spacecraft that looked like an airplane and could fly back to a runway was what we had all dreamed of while watching Saturday morning cartoons and movies about space travel. It was a leap of imagination, an incredible technological achievement, and perhaps a little bit of magic.

I had never really planned on entering the world of Mission Control, or NASA for that matter. From my earliest memories I was enamored with anything that flew. But in those early youthful years, there were no spacecraft, so I concentrated on airplanes. I built models that didn't fly, then I moved on to those that did, and eventually I was given the opportunity to work on (and fly) full-sized craft. Of course, the excitement that went along with the early space missions captured my attention as well, and my math educator father kept me engaged with NASA reports and information he received at his job in the state education department. I built model rockets, and I probably knew more about the space program than anyone else at school. If it flew, I could be counted on to talk about it endlessly.

Despite my immersion into space missions as a child, I really wanted to build and fly airplanes for a living. If you had asked those who knew me as a child, they all would have said that I was captivated with rocketry—and of course I'd go to NASA. But I was filled with stories of the great pilots and airplane designers, from the Wright brothers to those of modern times. I was in my junior year of college when a twist of fate redirected me to space, a chance coincidence of a local airplane company going bankrupt and NASA, looking to rebuild their corps of personnel as the Shuttle moved toward flight, putting out a call for interns and students. NASA's Johnson Space Center had been struck hard by involuntary reductions-in-force and layoffs after the Apollo program ended, but with a new program about to lift off, it was clear that staffing levels needed to rise.

I was fortunate to be selected as a Cooperative Education Student (co-op), and I would spend the next two years shuffling between Houston and Minnesota a few months at a time as I finished my degree and learned the ropes of Mission Operations. When at last I reached the end of my schooling, a permanent job was waiting for me if I wanted it—so I took the leap and moved to Texas. I considered a job working on Tomahawk cruise missiles in San Diego, as well as a position in Fort Worth, helping to test the F-16. Since I had learned scuba diving in high school, and had gone on to be a paid diver and instructor all the way through college, I even considered staying in the scuba diving world that I had come to know well. I enjoyed the technical aspects of diving, and the risks associated with working in an alien underwater environment, and there were exciting things to be done there. But I think it might have been that first day that I stepped into the cathedral of Mission Control that tipped the scales. It truly was a magical place, and I was getting in on the ground floor of the Space Shuttle program.

Never before had mankind flown anything with wings so fast or so high. Never before did we have a spacecraft that was capable of so many things. And to do all of those things, we needed people. Specialists in mission design, planning, and consumable analysis. Flight controllers with a deep understanding of systems operation integration and troubleshooting. And engineering leaders who could look at complex problems and situations and marshal the right troops to get to a sound solution. I was fortunate to be on the front end of the staffing ramp-up—not quite the first, but early on—so I was blessed with the opportunity to be trained by the remaining veterans who had worked on Gemini, Apollo, and Skylab. Very few of those folks' names were known to the general public—the heroes of Apollo were mostly anonymous. But some were familiar—and I was just as awed to be working for Gene Kranz, the veteran Apollo Flight Director who became legendary when the history of that program really hit the public eye, as I was to be working with veteran astronauts who walked on the moon.

It was inspiring to come to work and sit in the chairs that had been occupied by those quiet but dedicated men and women who had dared and risked so much to take humankind off the planet. It was humbling to realize just how much there was to learn. Our training was intense and continuous—those who wanted to stand above an already tall set of peers had to put in the extra work, studying at night, looking at systems and operations well outside their own assignment. It was an incredible time, really—no one tried to rise above the rest by trampling on others— we were all doing outstanding things, and it was simply a matter of challenging oneself to go even further than the rest. Everyone improved because we competed with ourselves—and there really were no losers.

Training materials, classes, and simulator lessons were being developed simultaneously with our quest for knowledge, so we all took those workbooks home, we all sat through developmental simulation sessions, and we sat through classes on how systems worked and how they were operated. Nothing was static. I remember taking home thick workbooks on orbital mechanics and how telemetry and command systems worked. So did everyone else, apparently, because I always felt I was trying to keep up with the "smart folks." I remember waiting to be issued my first headset so that I could go to the Control Center, find an empty back room, and listen in on simulations to get the rhythm of the place. Headsets must have been in short supply back then, because only those assigned to the next mission seemed to have them. I must have somehow amused the veterans who saw my enthusiasm, because they signed my paperwork to go check one out. From then on, I was working extra hours to make sure I was not only getting my work done but also spending more extra hours learning how America flew people in space.

The Space Shuttle program was huge—spread out across the country, and across the globe. Comprehending the magnitude of the entire program was probably beyond the ability of any one single person— and for everything you learned, you found out there were a half-dozen

other things that you didn't know. While a co-op student at NASA, I
returned to school for a few months and also went back to my job as a
diving instructor and technical diver in Minnesota. In that group, I had a
friend who bounced around from job to job in the Twin Cities. One day,
he came to work with a box of metal parts—he had gotten a part-time
job as a quality control inspector at a small mom-and-pop metal casting
company. Those parts were familiar to me; they were the left and right
halves of a Shuttle rotational hand controller, used in all the cockpits we
had—including the cockpits of the real Orbiters—plus all the simulators
and trainers that needed hand grips. Honeywell was responsible for the
completed hand controllers, but they had subcontracted the metal shells
to this small firm in Minnesota—and I had been training with them in
my hands down in Houston.

This realization of just how big the Shuttle program really was never
left me—and it inspired me to be the best I could be, because there
was no way I wanted to let down the hundreds of thousands—if not
millions—of people who eventually had a hand in the project. Later, as
a flight controller and Flight Director, I would take trips to visit manu-
facturing plants and design centers for meetings and evaluations on various aspects of the program, and these visits would again remind me just
how many people it took to make our job possible. It was amazing, it was
humbling—and it made one proud to be a part of it.

As I had mentioned, becoming a Shuttle Flight Director was not
something that I consciously aspired to when I was young, but it became
a goal once I understood the structure of the human spaceflight organization. Being chosen is something I described as the Mount Everest of
my career—a peak that appeared to be the tallest thing on the planet,
with no further goals beyond. But in truth, it was merely a launching pad
for almost twenty years in the center seat of Mission Control. A lot happened in those twenty years: we flew an entire program with the Russian *Mir* space station. We built the International Space Station (ISS). We
flew to the Hubble Space Telescope (HST) numerous times. And we lost
another Orbiter and its crew. There were times with lots of missions, and

times where it seemed we'd never get off the ground again. Politics ebbed and flowed around the program, making many wonder where we were going—if, in fact, we were going anywhere at all. But all that time, the work of the Flight Director was to build and fly missions—and to do that as safely as possible. No mission is ever truly safe, but we do everything in our power to identify and minimize risks, to mitigate those that can be dealt with, and then to accept those that cannot. Answers were rarely right or wrong, but often lived somewhere in that gray space where your best tools are honesty and a willingness to face your fears directly.

This history that you are about to read brings to life those days on console and in the meetings rooms, simulators, and training grounds where we lived for three decades. My goal is not to try and tell the entire story of the Space Shuttle program, but to give the reader a feel for what it was like to be inside the flight program—the months and years that went into preparing for missions, the weeks spent flying, and the aftermath, where we figured out what we had learned so that we could go back and do it all again. There are stories of remarkable technical insights and stories where no one really knew what was going on. Flying is deadly serious—but it is often wrapped in a smile, and there is much to laugh at. My stories are unique, as are the countless thousands that others can and will tell in their own books, their own memoirs. My reminiscences here don't tell the complete story of the Shuttle flight program; they are the view of one person who was fortunate enough to have a wide view of it. So think of this as one perspective: a view behind the curtain of life in Mission Control, a life spent with dedicated individuals striving to do something unique, blasting people off the face of the planet and bringing them back after they had achieved an important goal.

It has been seven years since I last plugged my headset into a console in Mission Control, and some memories are beginning to fade, so it is time that I commit them to history so that others can learn what I was privileged to learn in all those years. I know that there are many space enthusiasts who would give just about anything to have been only a tiny part of a single Shuttle mission—and my colleagues and I were fortunate to be doing those entire missions over and over again for decades. Yes, *privileged* is the word—privileged to be trusted with such enormous resources to accomplish lofty missions. Privileged to be able to work with so many people who were so much brighter than I was. Privileged to have been at the pointy end of the exploration spear for so many years. It was hard, but it was worth it.

My stories include the trials and tribulations of what it is like to run a complex space mission. But while I hope that you learn something from all these tales, my real hope is that you appreciate the tremendous efforts put into raising humankind's sights, from ground level to the stars, by the hundreds of thousands of men and women who served in the program through its many triumphs and its horrible tragedies.

The Shuttle program is over, and many of the details of how it worked are already lost. But future generations of technology will come along—they must come along—and our destiny to move off this planet will be achieved. I don't know by whom and I don't know when, but I trust that those future explorers will have their own stories, similar to those captured here. I hope my stories will entertain future generations with what it was like to push the path into the sky.

—Paul Dye
Dayton, Nevada

The Last Time

No two shifts in the Mission Control Center (MCC) are ever alike. The facility is the same, of course—the big, gray, windowless building sits on the Johnson Space Center campus in Texas, along with office buildings and other large gray buildings that house labs, vacuum chambers, and engineering research facilities. Just a stone's throw from Galveston Bay, this former cow pasture had been turned into America's premier center for space exploration back when John F. Kennedy decided to send the nation's heroes to the moon, and it has grown ever since.

Some days in Mission Control start slow and build. Other days, you come right out of the gate with a major problem, and you never slow down. And then there are the days when you simply don't know where you're headed: how quickly you'll be on track or how quickly it will all go to worms. As I got out of my car in the Flight Director parking spot right next to the door (one of the perks of the center seat), I sniffed at the humidity and looked at the sky, wondering if the low clouds would break up for some afternoon flying later in the day. You just never knew on the Gulf Coast. At any rate, that was for later—now it was time to go to work. I passed through the double doors that kept the heat and humidity out of the carefully controlled climate of MCC and took the elevator to the second floor. I passed by the coffee pot without stopping—there'd be time for that later. I liked to focus myself on getting the shift started first. I passed the mementos and pictures of thirty years of Shuttle flight as I walked down the hall, then scanned my badge to open the door into the front room itself. A few flight controllers looked up and nodded a silent welcome as I headed for the Flight Director console in the center of the room.

I dropped my briefcase, pulled my headset out of its bag, and plugged it into the keyset jack under the console. The keysets were our link to the outside world—an intercom panel that allowed us to monitor and/ or talk on any of thousands of various communications channels, called loops, that reached out around the world, and out into space. Each loop had a name and a purpose, and it was organized in a hierarchical structure that allowed specialists to have conversations among themselves while generalists could talk about the big picture. *Generally,* everyone involved in a mission was expected to monitor the air-to-ground communications channel—the talk between the crew in space and the CAPCOM (Capsule Communicator, the one person in the Control Center who did all the talking to the crew) on the ground—at all times. They were also expected to monitor the Flight loop, formally known as the Flight Director loop, which was the primary channel where everything was coordinated. The Flight Director was the owner of this loop, and it was how anyone needing to talk with Flight (the Flight Director's call sign) provided information, asked questions—or pleaded their case.

Communications on any of the loops followed a protocol like that used by pilots and air traffic controllers for generations. If you need to talk with someone, you punched the Talk button on your keyset for that loop, and then spoke their call sign, followed by yours. "CAPCOM, Flight" would be how the Flight Director got the CAPCOM's attention. The proper response would be "Flight, CAPCOM—go ahead." This let the caller know that the receiver was listening. Conversation could then follow as required. Once that link had been established, conversation often became much less formal, with discussion, questions, answers, and brainstorming-type dialogue. This mix of formal calls and informal conversation often seemed strange to newcomers, but once they got used to it, the rhythm became obvious. You had to get comfortable with the fact that thousands of people in the program could be listening anywhere in the world to what you were saying, of course. But once you got over the fear of making a mistake (which everyone did now and again), communication came naturally—to most. There was a small percentage of folks who never got comfortable, and they simply didn't make it as flight controllers.

Getting the rhythm of Control Center communications could take a little time, but once you did, you could listen to multiple conversations at once, and pop in and out of ongoing discussions with ease. Reading transcripts of a recorded keyset was often difficult, because you couldn't recognize the voices the way you could when listening to an actual recording. Keyset audio was usually recorded in the Control Center, and flight controllers often asked for audio tapes of their sessions so they could go back and reconstruct what they had done right, and what they had done wrong. It was far easier to do this with audio than by reading a transcript.

There were two keysets on each console, one for the person who was currently responsible and one for the person coming on to relieve them—or simply for an observer who dropped in for a while. It was considered bad form to walk up and start talking to someone on console without being plugged in; you never knew what was going on in their other ear. They might be listening to multiple conversations all at once, and chatting with them face-to-face could break their concentration on something important. So, if you were just dropping in to see what was going on, you always wanted to plug in and select the same loops they had up on their keyset.

As I plugged in, sat down, and selected the loops I needed for my shift on this morning, I was pleased to find it generally quiet. Quiet is good in our business—everyone is working to a plan, and if everything is going well, there is little need for conversation. There was the usual hum of conversation among senior flight controllers in the Shuttle and Space Station primary control rooms and their support folks sprinkled in small rooms around the building. But the Flight loop was quiet, and the crew wasn't talking much, indicating that they were busy working on the day's scheduled activities.

All Space Transportation System (STS) missions had a number, with the final flight being number 135. The STS-135 crew was busy moving cargo between the Space Shuttle Orbiter *Atlantis* and the International

Space Station (ISS) on this quiet morning. The hatches were open, and the two crews were busy working their nominal timelines when I got a call from our environmental and thermal flight controller, call sign EECOM, regarding a change in cabin pressure (dP/dT). EECOM sat one row ahead, and over to my right, against the wall, and when he called me, I looked over in his direction. The look on his face told me we had a problem. Before we were through, our world had changed—and the rest of the team had been drawn into the chase:

"Flight, EECOM—we're seeing a negative dP/dT, and the crew has a master alarm. Recommend the joint emergency egress cue card in the Joint Ops. Page 331 in Joint Ops."

I keyed my mic—"Copy EECOM. CAPCOM, let's get them in Joint Ops."

Joint Ops was the Joint Operations Checklist—a document that covered all the procedures that involved interaction between the Shuttle and ISS crews (and their respective control centers). The ISS used very little paper—most of their procedures were electronic. The Shuttle team, working with a vehicle whose computer systems were designed back in the 1970s, depended more on actual paper books. But for critical procedures—like a cabin leak—the necessary procedures were generally printed on cue cards—thick paper that could be slapped onto patches of Velcro on the walls and grabbed quickly when needed. Getting the crew in the Joint Ops procedure allowed everyone to know exactly what was happening, and what was going to happen next, regardless of whether they were looking at the procedure in paper form or on a computer screen.

"Atlantis, Houston . . . we see your cabin leak, and we want you on the emergency egress cue card in Joint Ops. That's page 331 in Joint Ops, if you don't have the card handy." This was my CAPCOM getting the crew headed in the direction we wanted them—even though we knew they were probably already there.

I needed to make sure that the ISS was going down the same path we were, so I keyed my mic on my loop once again—"Station Flight, Shuttle Flight, my loop."

The Space Station Flight Director was Courtenay McMillan, one of my colleagues from the Flight Director Office. Courtenay was one of those incredibly sharp engineers who grew up in the Space Station program, and she had a deep understanding of the systems as well as all the details of the international partnership. I was also certified as an ISS Flight Director and served in the position when needed—but I had nowhere near the depth of knowledge of the dedicated cadre of which Courtenay was a member. I was lucky to have her there, really running the show. While docked, the Orbiter systems and operations were mostly idling at a low level and the figurative (as well as the real) center of gravity was in ISS operations, led by ISS Flight.

For Shuttle missions to the ISS, I was always on the Shuttle side. But during routine orbit operations, Shuttle Flight Directors filled holes in the schedule to augment the folks who were dedicated to ISS, and (more importantly) to keep us up to speed on the Station and how its operations worked. The ISS Control Center was just out the door and down the hall from our Flight Control Room (FCR), in the "old" MCC building. It had been changed and refurbished, and was more modern than the "new" Control Center where we flew the Shuttle, but the operational concept was the same.

Technically—and officially—speaking, during joint Shuttle/ISS operations, the overall lead for the mission was the lead ISS Flight Director, and we worked under that person when the Shuttle was docked to the ISS. But realistically, we were equals, and masters of our own vehicles. Since we shared offices (in yet another building on the other side of the space center), staff meetings, and staffs, there were no mysteries in operating together—and we consulted regularly throughout the mission. In an emergency case like this, we needed to be in lockstep through the procedures and decision-making.

"*Go ahead Shuttle Flight—we see the dP/dT.*"

"Okay, we've got our crew on the cue card, and I expect everyone's headed for their respective vehicles. Is your crew in Joint Ops?"

"*Yup—we've got them on the card too.*" Courtenay was fast, and I figured they were already with us—but a voice call made sure.

"Okay."

"*Flight, EECOM. T-Rez looks like about six hours, we'll refine that.*"

"Copy EECOM. Let's see what happens when they get the hatch closed."

Just like that, a quiet morning in MCC was interrupted by a large cabin leak in the joint stack—*Atlantis* and the International Space Station had been docked together for several days on this last of all Shuttle missions. The crew was busy transferring cargo in both directions, and suddenly, their atmosphere was going somewhere—but we didn't know where—or why. The Joint Ops Checklist was used for malfunctions that could affect both the Station and *Atlantis*. Both crews had copies, as did all the Control Centers so that we could work together. T-Rez was a complicated number that was essentially a time that indicated how long you had to work the problem. The bigger the number, the better off you were; the smaller—the worse things looked.

Flight-controller software tools had changed a lot over the thirty years of the Shuttle program. In the first decade of the program, controllers had essentially what the folks flying Apollo had: software running in the Mission Operations Computer (MOC) down on the first floor. Equations for specific computations and calculations had to be written and coded months before a flight and inserted into the MOC's software build for a particular mission. Displays were defined using a pencil and archaic worksheets, then were sent to the software division that coded them up. You could tell the MOC to perform a calculation with a dedicated button on your console. Changes were rarely possible after dedicated flight training had started for a mission and, oftentimes, flights were made with less-than-safety-critical software and displays not working right. Off-line calculations were often the quickest and best way for flight controllers to work, using calculators and small minicomputers to do computations specific to their discipline. Sometimes, worksheets were developed and documented in close-by handbooks for each discipline. It wasn't until the mid-1980s that real computer workstations started appearing on various consoles to do the heavy lifting that simply couldn't be done in the MOC—and flight controller creativity was unleashed.

The EECOMs of the latter days of the program had many software tools at their disposal. Computations for things like leak rates and T-Rez were done continuously and displayed on graphical plots, and rudimentary artificial intelligence algorithms backed up decision-making on potential future scenarios. Dumb green consoles of the Apollo days gave way to the blue consoles of the modern era, but the consoles themselves were simply fancy shells to house the workstation computers that ran the software that gave controllers more information and better options for every failure case.

"Houston, Atlantis. Just to let you know, we're all headed to the Orbiter [Atlantis], *and we can confirm we're feeling the depress—I just had to clear my ears."*

Just like in an ascending airplane, the need to clear your ears was a sure sign of an atmospheric pressure change. The crew was letting us know that this wasn't just a malfunctioning instrument—they knew that the pressure was actually dropping.

"Roger Atlantis. We copy that you can confirm the leak, and you're headed to Shuttle."

That was from CAPCOM, seated to my right. She knew she didn't need to wait for my permission to answer a call like that.

My CAPCOM (Capsule Communicator), Shannon Lucid, was no stranger to spaceflight operations. Today, she was sitting on the ground on my right, as she had done so many times before. But this quiet, smiling grandmother was a veteran of over six months on the *Mir* space station— only the second American to live aboard *Mir* and, for decades, the holder of the spaceflight duration record for women. Shannon had served as my CAPCOM more times than I could count—many joked that we were like an old married couple, each of us knowing what the other was going to do in a situation. It was good to know that I could trust her to repeat exactly what the team needed as she carried out her job of being the only voice between the ground and the crew. Shannon had quietly disappeared from the Control Center a couple of years before, when she was assigned to be NASA's Chief Scientist, stationed at headquarters in Washington, DC. Like me, she found that being a full-time professional resident of the inner Beltway was not to her taste. So, despite the promotion, she campaigned for a short stint in the job, and then a return to Houston and Shuttle operations. Shannon knew how to be cheerful on the outside, even when she might not have felt that way inside, and she rarely let folks see her without a broad smile. It helped the team to have her on console.

"EECOM, Flight. Any changes in the leak rate yet?"

"No Flight. It seems steady, but we've revised the T-Rez to eight hours now—we just have more data. We really can't tell anything until they get the hatches closed."

"Okay, all flight controllers, red on the Flight loop, please!"

Calling for a color was a time-honored tradition in the MCC—it was a way to make sure that everyone was listening, or that everyone had reached a certain milestone in a review, or that folks had given a "Go" on a particular request. In this case, I wanted to make sure that everyone was with me on the Flight Director voice loop, so I asked for a red. In the old days, each controller had a hardware panel with three buttons—red, amber, and green. The buttons lit a light in a matrix on the Flight Director console with the appropriate color. In the last half of the Shuttle years, the hardwired lights had been replaced by digital displays. They performed the same function as the hardwired lights, but they operated over the internet on the controller's PC. It gave the same effect, but I always chuckled at how much programming it took to replace those simple buttons, lights, and wires. I had all reds in less than ten seconds—people were paying attention.

"Okay, everyone, it looks like we've got a fairly large leak. Not sure where it is yet, we'll wait until we get the hatches closed to see whose side it's on. If it's on the ISS, we'll see what they need for us to do to help; if it's on our side, ah—EECOM, can you give me an idea with this size T-Rez? Uh... what does that translate into for Shuttle time on orbit—if it's on our side?"

"I'll work on that, Flight."

"Okay, and FIDO, have we got any landing sites underneath us right now? I guess I'm really asking, let's see...I'd guess a couple of hours to undock and clear, then a reasonable deorbit prep. What have we got out a couple of revs?"

FIDO was our Flight Dynamics Officer—essentially our navigator in Mission Control. FIDOs kept track of trajectories, where *Atlantis* was, and where it was going. They also computed maneuver and deorbit burns—so they were the ones who could find us a landing site when we needed one. TIG (time of ignition), another of their responsibilities, was when we'd have to do the burn—in this case, to drop us out of orbit. A rev was a quick term for a "revolution about the earth"—essentially a single orbit, around the planet.

With one of the largest back rooms in the Control Center, FIDOs sat in the front row of the FCR, over on the left side. They had a lot of computing power at their disposal and spoke in a language of vectors and targets that few outside their discipline understood. In the front room, we looked at the back of their heads most of the time, but when it was important—like now—they would stand up and talk directly at me— even though they still talked over the loops so that everyone else around the world knew what they were telling me. An experienced FIDO, Dan, had been through drills a thousand times, and he knew that it was too early to get excited about something as routine as a cabin leak. He looked me in the eye and I could see a little smile in his eyes, as if to say, "Sure, I can tell you the trajectory score, but we know EECOM will solve this."

"Flight, FIDO. We've got two West Coast opportunities in a row, Northrup and Edwards on the first, Edwards on the second. They are about three- and-a-half to TIG, and then a rev later."

"FIDO, get me those exact times, and give me clocks to TIG—we'll figure an hour after TIG for touchdown. Why don't you give us clocks for both TIG and touchdown for both opportunities—we've got the room up there."

"Roger Flight."

Up on top of the big screens in the front of the Control Room, there were slots for about eight different clocks. One always displayed Greenwich Mean Time, while another showed the total Mission Elapsed Time (MET). The others could be configured as stop watches or timers, counting down to a specific event, or counting up from when something had happened. Putting a clock up helped everyone focus on time—the single absolutely unrenewable resource that we had. Time always marched forward, and countdown timers always counted to zero, and no amount of pleading could change that fact.

"Okay, folks, in case the leak is on our side, I am thinking that if it's stable, and we can get on the ground with an hour's reserve, we're going to be coming down. I just don't want to cut it too close, so EECOM, I am going to be looking at you for a recommendation on..."

I was interrupted before I could finish the thought.

"*Flight, EECOM. We just saw a big jump in the rate—I think the hatch is closed, and the leak is on our side—I'll get a number for you.*"

"CAPCOM, Flight. Let's see where they're at and..."

Once again, someone else was ahead of me, and this time it was the crew. They, of course, couldn't hear what we were saying on the Flight loop, and everyone was trained to instantly stop talking when the crew came on air-to-ground—they were the reason we were there, so they always had priority.

"*Houston, Atlantis, Space to Ground One. We've got everyone on our side, and closed the hatch, and I guess you can see that we're still leaking. We're going to the Orbit Pocket, page 4-3...*"

"Okay, CAPCOM—copy that, we're following them to the Pocket. EECOM—is that where you want 'em?"

"*Flight, EECOM. Yes Orbit Pocket, page 4-3.*"

I turned slightly toward Shannon and nodded my head—a quick signal for her to pass those words up to the crew. It was faster than me having to verbalize it.

"*Atlantis, Houston. We copy you in the Orbit Pocket, and we're looking at the leak on the ground!*"

The Orbit Pocket—short for Orbit Pocket Checklist—contained all the immediate response procedures for things in *Atlantis* that needed to be worked immediately. They were the quick response steps used to make the system safe. The full malfunction checklist procedures could be dealt with after you stopped a leak and regained control, or had dealt with whatever emergency arose. The Joint Ops book told both the ISS and Shuttle crews what to do to get to their own vehicles and isolate themselves, and each crew then went to their own detailed book to solve their own issues. Ours was the Orbit Pocket.

Things were tricky with two spacecrafts. The Station was full of atmosphere—lots more than we had on *Atlantis* because of the comparative volumes. When the hatch between the two vehicles was closed, the leak rate picked up on the Orbiter because there was a smaller volume to leak out . . . and it was real trouble.

Now it was the EECOM's turn to stand and face me. I could see that he wasn't happy—he knew that all the focus was on him, and what he and his team could do would determine which path we'd be taking in the next few minutes. He wasn't happy because his data weren't solid—this was going to be a judgment call based on changing rates, which was getting awfully close to the definition of guessing—something that flight controllers never want to do.

"Flight, EECOM. My preliminary estimate is six hours to eight psi—we're working to refine that—give me a couple of minutes."

"Copy EECOM. Let's get the numbers right, and I need to get a feeling of your confidence—is that leak going to be stable, or is it varying? I'm trying to figure out how much I can trust the time you're going to get, 'cause that's going to be really close getting all the way to the ground."

Atmosphere on a spacecraft is life—but not just for the astronauts. Humans can survive at reasonably low atmospheric pressures (usually measured in pounds per square inch, or psi), especially if you can bump up the percentage of oxygen. Half of Earth's atmospheric pressure is below you when you are at 17,500 feet in altitude, and people live at that elevation routinely. The top of Mount Everest is at just about a third of sea level pressure, and some hardy souls have climbed it without supplemental oxygen. So keeping the crew alive was not my most immediate concern—we could flow oxygen into the cabin for a while to keep them breathing. No, the bigger problem in this case was cooling the many critical avionics boxes in the vehicle. Computers are very good heaters—they use electricity to run, make calculations, and pass data back and forth, but most of the electricity they use is discarded as waste heat. Some of that heat goes into the water cooling system designed for the task, and some of it goes into the air. You need that air to cool the machines, or pretty soon you'll start losing the primary General Purpose Computers (GPCs) needed to fly the Orbiter, or the many Multiplexer/Demultiplexers (MDMs) that collect data from the various systems, transfer that data to the GPCs (and the ground), and send control commands back to the machines. Without GPCs and MDMs, you couldn't fly the vehicle—and without the various sensors, inertial measurement units, and display screens, you wouldn't know where you were. All these boxes depended on the air in the cabin to keep them cool (to a point), and the lower the pressure, the fewer the air molecules available to carry that heat away. Thinner air, poorer cooling. Reentering with a leak means that the closer you get to the ground, the thinner the air in the cabin and, therefore, the less efficient cooling would become.

The bottom line was that we might have five hours to get to the ground—but the absolutely riskiest part of that entry was going to be the last ten minutes if the pressure kept dropping and the cooling kept getting worse and worse. There were no points for second best in this game, and my controllers knew it. It did no good to separate from the station, go through entry, and then lose the vehicle in the last few minutes because we had nothing left to fly it. This wasn't going to be an easy call for my EECOM. I gave him a minute to watch his data, then pulsed him on the loop.

"EECOM, Flight. You seeing any progress on the Pocket?"

"Yes Flight. It looks like they've got the cabin pressure relief valves and the dump valves closed—let's get them to close the airlock—there are a couple of other steps, but that's our best bet."

Before I could answer, an interruption came from the row in front of EECOM. It was common for controllers to slip in an additional call if something was going up to the crew, and we appreciated it because it was more efficient to make one call rather than two.

"Flight, MMACS. If they've got a spare hand, you can have them get out the leak detector and get that fired up. It's in the IFM tool locker . . . that's MA9F."

The Mechanical Maintenance Arm and Crew Systems, or MMACS (pronounced "Max"), was a complicated position that handled all the Orbiter's mechanical systems as well as the robot arm, maintenance, and crew systems. A catch-all position, they were responsible for all tools and loose equipment in the Orbiter. Today's MMACS knew that if there was a leak to be found and patched, he would be the one responsible for helping the crew through that process, so he might as well get them started looking for equipment.

"Copy, MMACS. CAPCOM, make a note of that, but hold on."

I wanted to check on more details before bothering the crew again.

"EECOM, Flight. We don't want to open back up to ISS yet because...?"

"Flight, we need to isolate the airlock to see if that's our problem, and..."

"Oh, duh—yeah, right, got it—let's get the airlock closed up. CAP-COM, let's expedite that!"

The airlock in the Orbiter was not used as an airlock during ISS docking missions—it was, in fact, the Orbiter end of the tunnel that led to the Space Station. It was mounted in the Orbiter's payload bay, and was, therefore, a potential location for a leak—and we couldn't reopen the tunnel to the ISS (to give ourselves more time with the larger volume of air) while still checking it for leaks.

"And Flight, EECOM. Let's flow max N_2. Open up the N_2 Systems 1 and 2 to the cabin, if we could."

"CAPCOM, let's add that—N_2 Systems 1 and 2 valves to open, and expedite closing the airlock to see if that's our leak."

"Atlantis, Houston. We need you to flow max N_2, open up N_2 Systems 1 and 2 to feed the leak, and we'd like to get the airlock isolated ASAP to see if that's our leak."

"Copy Houston. Flow max N_2. Airlock hatch close is in work right now."

By opening the nitrogen valves, located on the ceiling of the Orbiter's Mid-deck, we would feed the leak to keep the pressure in the cabin from decaying to a level that might prove dangerous to the crew and equipment. We had lots of high-pressure nitrogen stored in bottles in the payload bay for just such an emergency. The Orbiter generally leaked a tiny amount on every flight, so we could make it up with the stored gas. Flowing maximum amounts of N_2 from both systems gave us the best chance for keeping up the pressure while we searched for the leak.

"Shuttle Flight, Station Flight. Your loop." Courtenay was listening in and watching our progress like a hawk.

"Go ahead, Station Flight."

"Yeah, we see the leak is on your side. We're standing by over here." There was little that the ISS could do for us now, and Courtenay knew it. They'd have to watch as our crew fought the leak with what little help we could give them from the ground.

"Yep, we're just about to close the hatch on the airlock. We'll let you know what we find. We've got some landing opportunities, but they are going to be close, and frankly I'm not sure we've got stability on the leak—just hold on a couple of minutes, we may be back."

This was my indication to the ISS Flight Director that we might not be able to stop our leak, and might not have a landing opportunity—and that our crew might just be coming back to the Station.

"Houston, Atlantis. Uhh, we've got the airlock hatch closed, and we still see the negative dP/dT. We're opening up the N_2 systems now."

"Copy Atlantis. Stand by." CAPCOM knew to give me a moment to talk with my team and have the crew stand by for whatever we could tell them.

"EECOM, have you got anything else up your sleeve?" I wanted to make sure we had covered all the bases so that we could understand our options.

"Uh, Flight, EECOM. We aren't seeing much of a change in the overall leak rate, but it is a little unsteady—we're having to sort of interpolate a little. The hole size is probably about a quarter inch. I think we should see if the crew can hear it, maybe get the leak detector out—see if they can pin it down."

"Okay, CAPCOM. Let's tell them that we are still seeing the leak, and see if they have any leads on where it might be—anything they can hear or see flowing in one direction. We think it's about a quarter inch."

"Copy, and Flight, do we want them to continue in the Pocket?" Shannon was always keeping track of what we had the crew doing...that was her job.

"EECOM?" With one word, I passed the question on to EECOM.

"No, Flight, we've pretty much got that done—uh, stand by." I realized that standby probably wasn't for me, but for EECOM's back room. I knew his loops were buzzing, and because this was the focus of the failure, he was going to stay busy until we sorted this out.

I turned slightly to my right. "CAPCOM, just have them search for the leak for now."

"Atlantis, Houston. We see that the leak is still there, we think it is about a quarter inch and, uh . . . we'd like to know if you can locate it by sound or anything?"

There was a slight pause as the crew digested this, then they came back. *"We'll check, Houston."*

Searching for leaks was always a long shot. A quarter-inch hole could be anywhere. It could be a micrometeoroid hit on the skin, but the crew couldn't see the skin, not directly—everything in the cabin was covered with close-out panels or switch panels or lockers. And the leak might not be a hole in the skin—it could be a plumbing fixture that suddenly let loose—but those were mostly on the aft bulkhead, again, hidden from view by close-outs—or worse, down in the Lower Equipment Bay (LEB). The LEB was not an easy place to reach if it hadn't already been opened up, and sending a crewmember down there was truly exploring the bowels of the ship with lots of plumbing, tanks, and pumps. The question in my mind was not just where the leak was—but did we have confidence that it would remain the same so we could make a dash for the ground?

"Flight, FIDO."

"Go ahead, FIDO."

"Flight, we've got the LSOs calling up Northrup and Edwards, but I really think it is going to be close to get the first opportunity. We've got to get undocked, and separated, then do the emergency deorbit—I'm guessing that if we started right now, we'd barely make it."

The LSOs were the Landing Support Officers—basically our interface with the various landing sites around the world. Back in the days of capsules that splashed down in the oceans, they worked with the Navy and other military services to request the support we needed at any time. Since we now landed on runways, they worked with the airfields that we might use, including Edwards Air Force Base in the high desert of California north of Los Angeles, and Northrup Strip, a dry gypsum lake bed located within the White Sands Missile Range in central New Mexico. They had direct communication lines to both, as well as many more places we might have to land. Until we had ruled out an option, we wanted everyone to be ready—just in case.

It was simply a matter of geometry (and geography), that you had a landing opportunity at White Sands a rev before you had one to Edwards—often you'd have White Sands on one, both of them on the next, and then Edwards alone on the third. A rev takes ninety minutes, and you can accomplish a lot in that time—enough to make a landing feasible versus dashing to the ground in hope that you'll get all the procedures done in time.

"Yeah, I see your point—I agree with you FIDO—I don't want to screw up the sep, or plume the station, or mess up the deorbit prep. Let's shoot for the second opportunity—that's just Edwards, right?"

"Yes Flight. Just Edwards."

The sep, or separation burn, was the maneuver that backed the Shuttle away from the ISS, slowly and carefully to make sure that it didn't bump into anything or disturb its attitude control system (which keeps the spacecraft pointed in the desired direction). Another concern was the possibility that the Shuttle jet plumes (or exhaust) were directed at a solar array, making it wave like a flag in a stiff breeze—a sure way to damage something. Pluming the station was one of the worst things we could do during an undocking, and rushing through the procedure was a great way to miss something and cause exactly that to occur. Never rush anything if you can possibly help it—a fundamental rule of flight operations.

I turned my attention back to EECOM. "Okay...uh, EECOM. How's that rate looking—can you get me to the ground if that's, say...four and a half hours away?"

EECOM looked at me as he juggled a pencil. I could see the pressure he was under—and I was the one causing it. *"Well, Flight, our latest calculation shows we're going to be down to...uh, Flight, it looks like that is going to be just about our limit—if we go to eight psi now, then start feeding the leak, we are only going to have about a fifteen-minute pad to the ground."*

One way to slow the leak was to decrease the pressure in the cabin voluntarily to its lowest practical limit—eight psi. This would still give the crew what they needed to breathe and maintain the capability to cool the minimum amount of equipment—but it added a lot of overhead to the deorbit procedures, and we'd be operating on the razor's edge of keeping the equipment in thermal bounds. I felt I needed to lead the "witness" a little—take some of the pressure off by stating the worst case myself.

"Okay, so EECOM. You're saying a variation of the leak of about five percent is going to make or break us—is that right?"

"Uh, yeah Flight. That's about right.

"What kind of variation have you been seeing in the leak rate so far?"

EECOM glanced back at his numbers, and I could tell by the way he cocked his head that his back room was feeding him more information over his headset. *"Well Flight, that's hard to say, we really haven't had the time to . . ."*

It was time to make a decision. "EECOM, we're really not going to have any more data, or time to look at it, before we have to call it. Give me what you've got—can you get me to the ground at five hours from now?"

"Uh, Flight, I don't think we can be sure of that—I just . . . no Flight, that's too close—we could make the first opportunity with some confidence . . ."

"Uh, thanks EECOM, but I think the first opportunity—I'm going to rule that out. There are just too many things we could screw up if we tried for it that fast. Maybe in the old days, but not coming off the Station. So, EECOM—you can't make the second opportunity?"

I could see the pressure release from his shoulders as I stated the blunt conclusion for him. *"No Flight, I don't think we can."*

I looked up at the clocks one more time. Barely thirty minutes had gone by since the first hint of the leak, and we had been going full blast since then—time compression made it seem like ten minutes. The crew was waiting for our thoughts and words. The rest of the team was looking at me. We had all practiced emergency deorbits more times than any of us could count. Many of my team had been around for well over a decade, flying Orbiters for real and in simulations at least once a week. They knew how to get it home if I asked them to do it—and if the leak would allow. All they needed was the word.

I stood up and stared straight at EECOM. "So, EECOM—are you recommending that we stay on the ISS, leave the crew there?"

"Flight, EECOM. Uh, that's correct Flight—from this console, we don't see a way to make a deorbit work in this case."

This last of all Shuttle missions was unique. Ever since the *Columbia* had broken up on entry, we had always maintained an Orbiter rescue capability—we always kept enough supplies on board the Station to keep a Shuttle crew alive until another Shuttle could be launched to retrieve them. The rescue mission was a "canned mission" developed by a team I had led. Every mission's preparation and planning flow was put together such that the mission could be accelerated, the payload removed (or flown, if appropriate), and the Shuttle launched with a minimal crew to leave room to bring back the stranded crewmembers that had been left on the ISS. The damaged Shuttle would have been long-since jettisoned—deorbited with the payload bay doors open, by remote control, so that it would break up over the ocean. The crew of the doomed Orbiter would wait patiently for rescue on board the ISS.

The problem was, this was the last Shuttle mission—there was no rescue Orbiter. There were no External Tanks left, no sets of flight-ready solid rocket motors. There had been a debate over whether or not this last mission should have been flown—some thought we should have saved that last set of boosters and tanks to rescue the previous mission—and then simply not use them. But that would have been a horrible waste—and we had another plan up our sleeves. The Soyuz vehicles that brought long-duration crews to and from the ISS had three seats, and they were all planned to be used. But…the ISS could survive with fewer crewmembers if several Soyuz vehicles were launched less than full—with empty seats. We could bring home one extra person every time—and accelerate the Soyuz launches to maybe three per year, rather than two. I was concerned for the four crewmembers on *Atlantis,* and I was thankful we did not have a full crew of seven aboard. There was no way we could bring a stranded crew of seven home in any reasonable amount of time…but we *could* bring back a reduced crew of, let's say, *four* in a little over a year. Some unlucky Shuttle crewmember would be setting a new record for spaceflight duration—but they'd be coming home. The Soyuz was old, but it was dirt-simple reliable. A cannonball shot back into the atmosphere, parachutes for the descent, and a sudden—but survivable—jolt at the end. It was a sure thing, and a certain way to get our crewmembers home. What I was looking at was a huge gamble—a gamble that the leak wouldn't increase. A gamble that we could keep cooling the avionics. A gamble that nothing else would go wrong as we accelerated our normal undocking and deorbit, usually a two-day job, and tried to do it in just three hours.

The odds against a successful deorbit were pretty large. And while I have no problem with calculated risks, I am not a gambler. The four crewmembers of *Atlantis* were folks I had known for years—decades in fact. The Orbiter was going to a museum after the flight. I could bring everyone home safe, or I could take a huge risk and maybe bring the machine back to be decommissioned for tourists to gawk at. That deorbit was so tempting—the ground so close in time. The leak rate was steady now—but could I trust that? It had changed several times a few minutes ago. What if it was a crack, a tear— and aerodynamic forces opened it up? What if the vibration of entry through Earth's atmosphere shook loose the leaking fitting? What if, what if...

It was time to formalize the decision. There were hundreds of people in the building listening to the Flight loop, and many more in offices at all levels who probably had their ears cocked to their squawk boxes. Nothing that happened on the Flight loop was private, and the number of folks listening had probably increased as soon as the first person outside the Control Center heard the words "cabin leak" and told someone else.

I sat up straight and keyed my mic. "Okay, everybody, amber on the Flight loop, and Station Flight, Shuttle Flight—are you up on my loop?"

"I'm here Shuttle Flight."

"Okay, folks, I've got everyone...I think we've pushed this as far as we can go—FIDO, you're still not comfortable with the first opportunity, are you? I'll tell you, I'm not."

FIDO came back quickly. *"No Flight, I don't think that's reasonable."*

"Okay, then—we're going to have to go to lifeboat mode—ISS Flight, are you ready for some long-term guests?"

"Uh, yes, Shuttle Flight, we'll make room."

It was time to tell the crew of the decision that they probably knew was coming. Shuttle Commander Chris Ferguson, "Fergy" to just about anyone who worked with him, needed to be told directly what the situation demanded, and what we were going to do. I wanted the next call to go directly to him to make it official.

"Okay, CAPCOM, let's tell Fergy that it's time to reopen the hatches and start to gather equipment—uh, ACO, have we got an equipment list for stripping the Orbiter?"

The Assembly and Checkout Officer (ACO), one row ahead of me and to my left, responded. *"Yes, Flight. Page 4-1, in the Stowage Checklist."*

"Okay, CAPCOM, let's get them stripping the ship and retreating to the ISS."

ACO, in our control room, kept detailed records of every item in both the Shuttle and ISS—where it all was, what it was—and most importantly, what the crew needed to take with them in the case they had to abandon ship. That list was going to be needed—right now.

"Atlantis, Houston for Fergy."

"Go ahead, Houston, this is Fergy."

Shannon keyed her mic again. *"Fergy, we've been watching the leak, and the next landing opportunity just looks like we are going to be cutting it too thin. We are going to have you reopen the hatches and strip the Orbiter—we're going to have you retreat to the ISS and stay there. We're going to abandon Atlantis. You can find the list of gather hardware in . . ."*

I tuned out and sat back in my chair—the hard work was done—the decision was made, now it was just execution. We had plenty of time to get whatever the crew needed or wanted to strip from the Orbiter—food, clothing, spare parts, tools, Launch and Entry Suits, seats. The Station had lots of air to feed the leak for the hour or two it would take to render the vehicle derelict. Then we'd set it up for deorbit, install a set of jumper cables that allowed us to fire the orbital maneuvering system engines (by remote control from the ground), and send her to a fiery end over the Pacific. We had to do this all before it ran out of air completely—because once again, we had to do it before the computers overheated and we lost control . . .

But that would be for the oncoming team—I'd already sent an email to get the ball rolling and get another team into MCC to help my team make sure we didn't drop the ball—it would be terrible to drop an

Orbiter on a land mass. We'd done what we could, we'd kept the crew alive and safe. The goal of the Flight Director is twofold—crew safety and mission success. You couldn't have a successful mission if everyone didn't come home, and of the two goals, crew safety was first. It would be a year before I saw Fergy back on Earth, but at least I knew he—and the rest of the crew—would be alive, if albeit a little worse for the wear.

"Flight, Sim Sup on the Flight loop!"

Sim Sup (pronounced "Sim Soup")—the Simulation Supervisor—was a voice I hadn't heard for about an hour—ever since this whole thing got started. She and I had a private communication loop that no one else could hear, but I hadn't heard a peep from her end as we worked through this problem.

"Go ahead, Sup. Ahh... Station Flight, are you here too?" No need to make Sup's counterpart on the ISS side tell the same story.

"Yes Shuttle Flight, the Station team is here."

"Flight, Sim Sup, that was a good run, we think you came to where we thought we wanted you to be, and when do you want to debrief?"

"Let's make it ten minutes, Sup—I think we can use a quick break before we talk about it... so everyone be back in ten, and we'll debrief."

And just like that, we all leaned back from our consoles, took a deep breath, and realized that once again, it was all just make-believe. The Sim Sup had scripted a clever problem—but no one was going to be at actual risk. The *Atlantis* "crew" (Fergy, Rex, Sandy, and Doug) were safely on the ground, just across the duck ponds from where we were sitting. This was the last run of the last simulation for the orbit teams of the Space Shuttle program. And once again, like countless times before, we'd worked our way to a point where we could bring our crew home.

The art of any simulation is making it as realistic as possible. We like to say that we fly the way we train—but it is equally important to train the way you fly. One instantly unacceptable answer in any conversation on the Flight Director loop is *"Well, Flight, if this were a real day..."* That would get you slapped down pretty quick, and it doesn't make any difference who is sitting in the Flight Director chair at the moment. Fly the way you train, and train

the way you fly—and always, take it seriously. I always told my flight controllers that if they didn't leave a sim exhausted, then they weren't really putting
their heart into it. Thirty-three years in Mission Control took a lot of heart—
and it was only in retrospect that I realized just how tired it made us all.

The last time I sat in the Shuttle control room was shortly after the (real)
landing of STS-135. I had released my team—the planning team—for the
last time about ten hours before, making way for the entry team who
handled the last shift of the Shuttle program. The entry went smoothly,
as expected, and the postlanding handover to the Kennedy Space Center
took its normal course—everyone usually wished that part would go by
faster because our job was done, but little things always slowed it down.
No one really cared that it ran long that day; no one really wanted to leave.
I watched the landing from the Director's Suite, which was a little office
space above, behind, and to the right of the Flight Control Room (FCR)
with two big windows looking out over the floor. The suite was allegedly
designed for the Center Director to use, but in actuality it was the gath-
ering place for off-duty Flight Directors, former Flight Directors, managers, senior astronauts—anyone who wanted to feel like part of the action
yet had no reason to be in the FCR itself. The crowd in the suite grew
and dwindled from flight to flight depending on what was going on, and
what time of day it was happening. Ascent and entry were the big times,
of course. Very few folks spent any time in the room during orbit operations, unless something very unusual was going on. But, it was a good
place to informally catch up on gossip with other senior members of the
team before and after mission management team meetings. Most everyone who had access to the room had been in the trenches themselves, most
for many years. There was really no rank there; you spoke your mind, told
jokes, even heckled the team a bit if things were going well. They couldn't
hear you, of course, but it relieved the tension and made everyone think
they were still part of the game. Weather decisions received the most comment, usually—weather was always a matter of opinion.

But this last day—the last time we'd sit in the room during a Shuttle operation—we all sat and watched, made a few comments, let our breath out collectively at wheels' stop (the moment the Orbiter stopped moving on the runway, and the mission was over). We waited for handover and the news that the doors of the FCR were unlocked. What followed was the largest and longest group hug I'd ever been a part of. The room was literally flooded with people—they came in waves—mostly flight controllers and operations people who just wanted one last touch of what it had been like to be a part of the program. Older Flight Directors and astronauts, senior management who had risen up through the ranks—there was barely room to stand, and plenty of noise to go around. The Shuttle FCR was now just a room full of people shaking hands—and saying their goodbyes.

It was actually early in the morning Houston time. Landing was just before sunrise at the Cape, and handover an hour after that, so the sun was just reaching Houston—yet so many came to say goodbye. I hadn't bothered to go home after my shift. I just took a nap on the couch in the suite until a few folks showed up for the initial stages of deorbit. The post-handover backslapping and hugging reminded me of the last episode of the old *Mary Tyler Moore Show*, where the characters couldn't seem to break the bonds of a final goodbye hug and finally shuffled out of the room together, in a large embrace. In the same manner, it took a long time before people started to leave the FCR. At least an hour passed before gaps appeared in the crowd. Some decided to head off to breakfast, some headed home to bed. A few of us simply couldn't get it in our heads to leave. We wanted to be the last ones there.

In the end, there were about six of us. The viewing room was empty by this time. No one was coming back in, and other than Flight Directors, the ground controller was the only person left. We had been told that the Flight Director's flight data file—the exact duplicate set of all the books and checklists that the crew had on board—was not going to be inventoried after the mission. It was never going to be used again, so they didn't have to prepare it for the next flight. It was, in fact, surplus. So we split it up. Being the senior orbit Flight Director (by an unfortunately wide margin), I took the orbit checklists. The Shuttle lead for the mission hung on to his Rendezvous Checklist. The ascent and entry guys took their books as well. They were just checklists that we all had electronically, but there was something magic about these paper copies—and we knew that if we didn't take them they would be headed for the trash. We hung on to them because they had been an important part of our lives for so long. They were mementos, as surely as the patches, pins, and stickers we all had overflowing from our desk drawers.

After we divided up the spoils, we sat back and realized that we simply had no place else to go—nothing left to do. Sure, we'd all be back the next day, or the day after that, working in the office, looking at work that needed to be done for the Commercial Crew program, or the Orion program, or the ISS program, itself. If you wanted to remain in "the office" (as we referred to the Flight Director Office), it was a given that you had to get or stay certified for Space Station console duties. I had been an ISS Flight Director on and off since the program's beginning. And I had picked it up again in earnest a year or two before we finished with the Shuttle. So we'd all be back in the same building—but not flying a winged spacecraft. We sat there and exchanged stories, and at some point, we left as well. But not before a cooler with a few beers arrived. Drinking? In Mission Control? It seemed like a good idea at the time, but boy, was it nonregulation. I don't remember who was last out the door, but I do remember looking back at a very quiet place. The FCR was rarely noisy—conversations into headsets rarely rose above a low murmur, but this silence was the silence of a weekend holiday when nothing was in the air. It was the silence of abandonment, and the moment struck me as the same one that we had faced on that last simulation case: the final decision had been made and there was no going back. An era was ending, a glorious era. The only difference was that while the decision to abandon ship during that last training case was the right one, the decision to leave the Shuttle program behind was very hard to justify for anyone with the spirit of exploration running in their blood.

Chapter 1

The Most Complex Flying Machine Ever Built

The Space Shuttle program was, to make the understatement of the century, huge. Five space-qualified vehicles flew a total of 135 times. Originally approved in the early 1970s, the Shuttle first flew into orbit in April 1981 and landed for the last time in July 2011. Before then, the original atmospheric flight vehicle (*Enterprise*) dropped off the back of the 747 carrier aircraft for four free flights, landing by itself on the dry lake bed at Edwards Air Force Base in the Mojave Desert. Literally millions of people had a hand in this multigenerational project over its lifetime. People came to the program from all walks of life. From the engineers who designed it to the craftsmen who built it, from the planners and program managers who kept track of every little detail to the pilots and flight controllers who executed the missions—it was an all-consuming way of life for those involved at just about any level.

It was easy to forget that, near the end of the program, we had people working in key positions who weren't even born when the Shuttle flew into space for the first time. In the long run, of the six airframes built, two were lost in tragic accidents and four survived to be placed in museums long before their time. The same two launch pads were used at the Kennedy Space Center (KSC) from start to finish, and all the vehicle prep was done nearby in the Orbit Processing Facility and the Vertical Assembly Building. Thousands of people worked tirelessly at KSC to get the birds ready to go, and to receive them when they came back. Space Shuttle program management, the operations team, and the astronauts all lived in Houston, working at the Johnson Space Center (JSC). Mission Control was located there, in the same building it occupied for the Gemini and Apollo missions that preceded the Shuttle. The building grew over time, with flight controllers and ground teams eventually occupying five different control rooms and many more support rooms throughout the life of the program.

The JSC Mission Operations team, as it came to be called, was responsible for planning Shuttle missions, training the astronauts and flight controllers for those missions, and then conducting those missions, from the moment of liftoff until the vehicle had been safely returned to the runway and turned back over to the KSC team, who would then prepare it for the next mission. Mission Control was the focal point for the entire operation when a Shuttle was in space, and the center of that focus was the Mission Operations Control Room (MOCR), which oftentimes was referred to by its simpler name, the Flight Control Room (FCR). Teams of a dozen or more flight controllers resided in the FCR at all times during a Shuttle mission, led by a Flight Director. But the FCR flight controllers were really just the tip of the iceberg—each of them was supported by a backroom team of between two and ten people, and those support rooms were supported by their own discipline teams. In addition, the entire engineering community was available and participated in troubleshooting and evaluation work throughout a mission. So when people around the world saw the FCR, what they saw was a fraction of the staff. They really had no idea the number of people supporting those men and women who oversaw the entirety of a Shuttle flight.

Sitting in the middle of the FCR was the Flight Director who oversaw all real-time operations and, in fact, had the responsibility and authority to do whatever he or she felt was necessary for the safe and successful execution of the mission using procedures and rules that had been developed over the years of space operations. A Flight Director led the team to execute the mission plan and to modify it as required to achieve the mission they were sent to do and then bring everyone back to Earth. It was easy for those outside the business to think that the astronauts and the people in that small room did it all, but anyone who filled one of those positions would be the first to admit that we all stood on the shoulders of giants—and an uncountable number of giants at that. It was a huge operation, and it was hard for those deep in the trenches to get a handle on it. It was the job of the Flight Director to "conduct the orchestra," so to speak, to bring all the disparate parts, plans, and systems together to achieve the mission goals. The hardware and software (supplied by the engineering and programming community) had to be ready, the mission planning had to be ready, the payload had to be ready, the crew and flight control teams had to be ready—and when they were, it was a magnificent thing to watch. Many would argue that the team was at its best when it encountered difficulties, and it would be hard to argue with that—for watching great people use their skills and talents to overcome obstacles is an amazing thing to behold.

The Space Shuttle was the product of many different disciplines working on its physics, structure, and systems. A Flight Director had to grasp it all—though he or she was surrounded by experts who could support that knowledge with details. And the Shuttle was all about details. Millions of lines of computer code had to be right. The mathematics and physics of orbital mechanics had to be calculated to many significant digits. Systems were only as good as the tiniest nut or bolt. And everything depended on the skills and knowledge of the many people who had to do complex jobs accurately and at exactly the right time.

Before we can really talk about what it was like to fly the Shuttle, we need to understand how and why it flew. From the basics of orbital mechanics to the esoteric nature of complex systems and flight plans— it's important to know why and how things worked on the Shuttle in order to put the events during thirty years of missions in context. New flight controllers spent at least two years coming up to speed on the basics of spaceflight before becoming certified for the simplest backroom position. I'll try to condense that primer down a bit—but still give you insight and an appreciation for the complexities of the Space Shuttle systems.

To many, the Space Shuttle looked like an airplane. In fact, that observation isn't far from the truth. Built primarily of aluminum and some exotic metals, the Shuttle was manufactured by an airplane company (North American Rockwell) using the knowledge and skills they had obtained from decades of building some of the finest aircraft ever to grace a military ramp. Immediately preceding the Shuttle program, North American built the Apollo Command and Service Modules, and before that the X-15 research aircraft. Before that—some of the best fighters and bombers of World War II and the Cold War. They knew how to build airplanes.

The Orbiter itself was the size of a small airliner, a short DC-9 perhaps. Because I grew up with the early years of the space program, it was easy for me to think of spacecraft being the size of those used during the Mercury, Gemini, and Apollo programs. So when I first saw an Orbiter, I was shocked and amazed—the fact that something this big could go into space was almost incomprehensible. You almost have to look at the two generations of spacecraft side by side to see just what a huge leap the Shuttle was from what had come before. It wasn't as tall as a Saturn V, the rocket that took Apollo to the moon—but so little of that vehicle came back. The fact that the entire Orbiter returned was simply mind boggling.

The Orbiter had a fairly conventional configuration for a delta-winged aircraft (one with all the lifting surface located all the way aft, and shaped like triangle). Its fuselage was taken up mainly by the cargo bay, the crew compartment was at the front, the wing was fairly far back, and it was topped off with a tall vertical tail. There was no need for a separate horizontal tail because of the delta wing, and both pitch and roll control were provided by the elevons—a combination of elevators and ailerons. The rudder (the hinged rear portion of the vertical tail), was designed in two pieces and hinged at the front. These two sides could split, spreading out in a V shape to act as a speed brake for better management of energy during the final approach to landing.

The aircraft had conventional tricycle landing gear, with two wheels on each main gear strut, and twin nosewheels on a single strut. Many airplane-savvy people who saw an Orbiter up close were surprised to find just how far aft the wings were mounted—but that was a consequence of the delta planform. The main spar (the primary load-carrying structure in the wing) was actually located at the very rear of the payload bay, so most of what people considered to be the usable space in the Orbiter was located ahead of the wing. This actually created challenges when designing missions because you had to work hard not to make the vehicle nose heavy by putting too much stuff up front. The ideal payload was a useful dense mass in a small package, located near the rear of the payload bay—anything heavy and forward was just a problem.

The payload bay was mostly empty space, of course, until it was loaded. Two large payload bay doors allowed the entire top to open up, and the Shuttle always had to fly with the doors open when it was on orbit to expose the cooling radiators (located on the sides of the doors) to dissipate the heat generated by the vehicle's systems. The crew compartment, located forward of the payload bay, was the only pressurized, habitable volume in the spacecraft—the payload bay and the aft compartment (which could be considered as the engine room) was allowed to operate at a vacuum while the vehicle was in space. A series of vents allowed air to flow out of the unpressurized spaces as the vehicle flew out of the atmosphere on ascent, and then flow back in as the vehicle descended to land. Mechanical doors closed the vents off during the hot part of entry to prevent high-energy plasma from torching anything inside.

The engine room, or aft compartment, housed all the mechanical systems needed to support the main engines. Their power heads were actually inside the structure. The main engine nozzles stuck out the rear of the compartment, of course. The aft compartment itself was filled with plumbing and wiring—the guts of a mechanical beast, with all sorts of tubes, ducts, and pumps representing the mechanical organs of the monster. Although the types and functions of the equipment might baffle the average aircraft mechanic, the construction and complicated nature of wiring and plumbing equipment would have made him feel right at home.

Attached to the upper rear corners of the aft fuselage were two large pods—one on either side—that housed the Orbiter's aft maneuvering jets: the Reaction Control System (RCS) and the Orbital Maneuvering System (OMS) engines. All the propellant for these rockets was stored in these aft pods. Along with another set of tanks and jets in the nose (the forward RCS pod), these accounted for all the fuel used to maneuver the Orbiter while it was in space.

Fuel and oxidizer could be shared between the right and left rear pods, but the forward pod was completely separate, as no one wanted to run propellant lines all the way down the length of the vehicle. The more fuel line you run, the more chances you have for leaks, and the propellants used in the Shuttle were pretty volatile. So the pods were designed to be isolated front to back, and that inability to share propellant between all three pods caused some challenges for the flight control teams over the years. It was manageable, just another limitation, but it would have been so nice at times to be able to share gas all the way around.

As far as the crew compartment, the space available for humans was remarkably large when you compared it to previous spacecraft. Going from a Mercury to a Shuttle would have been like going from a broom closet to a living room—and the Shuttle even made the Apollo feel like a small bedroom. The crew compartment was divided officially into three decks, although this confused the heck out of most people because only two were actually usable. The upper deck, the Flight Deck, was where the primary flight controls resided, and where you really flew the vehicle. The Mid-deck was really the living quarters. It's where everything was stowed, where you could eat and sleep, and where experiments were usually stowed and operated until larger payload-bay-mounted laboratories came long. We thought of these two spaces as the upper and lower decks. But the Mid-deck was given its name because the Lower Equipment Bay (LEB) was located below the Mid-deck. Technically, yes, it was another deck—but it wasn't a basement. It was more of a crawlspace, and it was loaded with tanks, plumbing, fans, pumps, and all sorts of mechanical gizmos that needed to operate inside a pressurized space (unlike all the gizmos located in the vacuum of the aft compartment).

The crew compartment also had an airlock—used to prepare for space walks (aka Extravehicular Activity, or EVA) and as a tunnel connecting to the payload bay. It was located in the Mid-deck, and it took up a fair amount of space. Later in the program it was moved out into the payload bay to serve as a mount for the docking system used to connect to space stations such as *Mir* and the International Space Station (ISS). Moving the airlock gave the crew a lot more room in their compartment, and while it took room from the payload bay, the fact that payloads rarely took up all the space available made this an easy move. The Mid-deck was famous for its walls of lockers that lined the forward and aft faces. A Mid-deck locker was about the size of a small suitcase, and could carry 60 pounds of stuff—whatever that stuff might be. There was a repeating pattern of attachment points on the walls where they could be mounted—just like a set of LEGO bricks. It was a flexible and useful system that allowed for prepacking of cargo and consumables (like food and clothing) from anywhere in the world. Lockers could be shipped fully loaded to the launch site and installed in the vehicle, meaning that no one on the pad needed to pack individual items.

The lockers started out being identical, but it wasn't long before someone invented the double locker (it mounted in the same space as two regular lockers). And then someone incorporated an experiment into a locker-like device, with a control panel on its front door. And then someone invented a collapsible locker, and a soft locker. But everything was always referenced back to a standard size and weight—the Mid-deck Locker Equivalent (MLE). Hundreds of years from now, when humans are exploring distant planets in spacecraft of unimaginable size and complexity, my bet is that they will still be building cargo and equipment manifests based on MLEs, and no one will remember where the term came from—just like today when we measure liquids in gallons.

The Mid-deck was the location for sleeping, eating, and, of course, using the bathroom—the Waste Management Compartment was located over in the corner, just near the side hatch, where the crew entered the vehicle. The Waste Management Compartment was a little alcove that housed the space toilet and lots of hygiene stowage—toilet paper, towels, cleaning supplies, and the like. Famous for malfunctions, the toilet actually worked pretty well most of the time—it is just remembered for the times it failed because, well…it was more than a little inconvenient to have a crew of seven people all floating around, waiting for the toilet to get fixed. Very few things—other than maybe a fire or a cabin leak—took precedence over fixing the facilities when they were having issues. Thousands of items were stowed in the Mid-deck, from food and clothing to cameras and scientific equipment. Tools for fixing things, spare parts for the Orbiter and space suits, and all the accouterments for daily living filled the locker space. And bulky items such as Launch and Entry Suits (the orange suits the crew wore just for getting into orbit and coming home) took up nooks and crannies in the corners.

The Flight Deck was usually crowded with people, of course—that was where so much of the work was done when you were flying. The robot arm was operated from there, and windows looked out on the payload bay to provide line of sight for moving things around. The forward part of the Flight Deck was familiar to anyone who has ever seen an airliner's cockpit—lots of switches, instruments, and control panels all around. All critical controls were placed here, things that needed to be operated in a hurry on ascent or entry by the commander and pilot. The aft station controls were for on-orbit operation. Located on each side of the aft station were consoles that could hold mission-specific panels—controls that were designed to operate specific payloads. They would be changed out to support whatever mission was being flown.

While public relations pictures of the Orbiter's Flight Deck usually look pretty clean and organized, the reality is that when a Shuttle was flying, the Flight Deck became a sea of wires and cables fastened to console and panel surfaces with Velcro straps and loops of duct tape. Cables were needed to connect laptop computers (which didn't exist when the Orbiter was designed, so there were no provisions for LAN or power cables), for cameras and camcorders, and for all sorts of experiment cables that needed to connect boxes to each other and into the Orbiter systems for power and data.

In the early Shuttle days, we envisioned that it would be far too complicated to have every astronaut wearing a headset with a cable attached to the vehicle—everyone would get hopelessly tangled up, or would have to remain in one spot for hours. So we designed a wireless communication headset—a box was worn on the hip or thigh, and a headset was connected to that. It was an early form of a cell system, in some respects. And it was awful. A lot of time was spent trying to connect up, there was interference, batteries went dead...I remember long training sessions with multiple crewmembers all trying to figure out how to make it work. Eventually, the system simply fell into disuse in favor of a simple handheld microphone and a cabin speaker—just like pilots used in the old days, before the invention of comfortable, lightweight headsets. We discovered that it was just as easy to let everyone hear the ground calling, and for someone to pick up the mic to answer back. It was only during ascent and entry that you needed instant communications anyway—going simpler just made things more civilized on orbit.

While it is easy to concentrate on the form and structure of the Orbiter when looking at how it operates, the real heart of the vehicle were the systems that made it work. The early flying machines rarely had anything more than cable-controlled flying surfaces and an engine that ran at full throttle (when it ran at all). The pilot had a wicker chair and maybe a rope to use as a seat belt. There were precious few instruments—maybe a couple of gauges to monitor the engine—and, of course no communications other than shouting at people on the ground.

Modern aircraft have a huge variety of systems, many of which now challenge the Space Shuttle in complexity and sophistication. After all, the Shuttle design was essentially frozen in the mid-1970s, and a lot has happened since then. But even given that modern technology has progressed, the Shuttle is still one of the most complex flying machines ever built, primarily because it had to be a rocket, a space station, and a winged entry vehicle—not to mention an airplane—all in one package. As such, the variety and complexity of the systems were far beyond just about anything mankind has ever built. In order to understand how and why the Shuttle was operated and flown the way it was, it is necessary to take a look at those systems and the people who operated them—so let's begin.

On the ground in Houston, down near the front of the MOCR, located close to the trajectory folks, you could always find a pair of flight controllers who worked very closely with each other and with the folks who were figuring out where the Shuttle needed to go. These two controllers were in the middle of things because they were in charge of the systems that did the onboard navigating, plus the jets and rocket engines that made it go. The first of the two, the Guidance, Navigation, and Control (GNC) Officer, looked after the inertial measurement units that tracked where the vehicle was in space, as well as the sensors that fed the Inertial Measurement Units (IMUs) with information on the outside world—things like star trackers, accelerometers, air data probes, and the like. The GNC Officers lived in a world of precision measurement, and their software was as complex as their hardware.

The Shuttle knew where it was because we started out by telling it exactly where it was when it started. It kept track of its location by measuring acceleration in three dimensions, and then adding each bit of motion to its last known location to come up with its new position. It's really no more complex than knowing that you started at your front door, moved 20 feet north and 5 feet west, then climbed 5 feet up a ladder that was leaning against your old oak tree. You know exactly where you are relative to the front door of your house—and you better keep your balance because that ladder looks a bit rickety. Tracking a vehicle in space is, of course, more complex than that because the equations of motion are affected by gravity. But, essentially, you are doing the same thing.

The Shuttle measured accelerations using finely tuned accelerometers that were strapped to the IMU stable platform. It was stabilized by spinning gyroscopes (in the early days), and in later years by solid-state devices—similar to what you have in your cell phone today. Picking a reference frame that is stationary among the stars is surprisingly straightforward. Earth may be rotating underneath you, and it may be going around the sun, but you can compute those motions, knowing Newton's laws. You can also compute your own motion because you are measuring it relative to those same stars. For the Shuttle, this was done using the IMU and accelerometers mounted on its stable platform. Just like measuring your motion in the front yard, we were able to tell exactly where the Shuttle was, so long as we knew where it had been.

So how did you know that the IMU platforms were properly aligned to the stars? Well, you took a look at the stars every once in a while using automated star trackers—you can think of them as specialized cameras that were bolted to the same solid piece of metal to which the IMUs were mounted. The Shuttle had two star trackers, and each of them could track stars by measuring where they were in the field of view and how bright they were. If you started out assuming that you knew how the platform was pointed, you could tell the vehicle to put a particular star, say, Polaris (the North Star) in one star tracker and Sirius (the brightest star in the sky) in another. The Orbiter would maneuver to the necessary attitude and, hopefully, the two stars would be located approximately where they were supposed to be. Now nothing is perfect, and errors can drift into your knowledge of how you are pointed. So what the trackers did was measure the error between where the stars were supposed to be in their fields of view and where they actually appeared. They did this with great precision. With two stars in the tracker, the computers could compute the platform's alignment, and the crew could order the computer to take that error into account, realigning the platform and once again giving you the necessary information to track your progress through the sky.

What happened if you lost the platform alignment completely? What if the IMUs tumbled, or the computers really messed up and lost their knowledge of how you were pointed? Well, then it was time to get out the star charts and a special gunsight-like device known as the Crew Optical Alignment Sight (COAS). Pilot astronauts trained to identify stars, then maneuver to put the stars in the site, and tell the computer which star they were pointing at. The computers would mark that position. Then they'd maneuver to another star and mark that position. Since the computer knew the positions of the stars, and you had now told it where the platform was relative to those stars, it could once again figure out its attitude relative to the stellar reference frame and once again track its motion.

So attitude was important—but so was knowing the components of your velocity (which were derived from accelerations) so that you could add up the motion in each of the three dimensions. We did this by starting with a special set of numbers known as a state vector. The state vector, in simple terms, was your position and velocity in space, in three dimensions, at a specific time. If you know the six components of position and velocity, and the time you were there, you can add up all the sensed motions since that time and figure out exactly where you were at any given later time. It is an elegant solution, and not that different from navigating on the surface of the planet, keeping track of where you have been. The scheme just does it in three dimensions. Of course, no man-made sensor is perfectly accurate. The longer it has been since the time of your state vector, the more error there was bound to be in your position. Fortunately, the ground kept track of the exact position using tracking stations and satellites, and every so often we'd send a new state vector up to the computers on the Shuttle, starting the whole propagation process over.

Early in the Shuttle program, state vector updates came a couple of times a day. But sensor accuracy improved over the years, along with the equations of motion. In later years, you could go much longer between state vector updates or alignments—this was a good thing, because once you were docked to a space station it was much harder to maneuver to find stars and check the alignment.

Now, of course none of this was possible if you couldn't maneuver, and doing that was achieved with the jets and rockets that were looked after by the Propulsion Systems Officer (PROP). PROP was responsible for all the thrusters, the fuel and oxidizer tanks, and the complex of pipes and valves that connected them all together. These thrusters came in three basic sizes: small, medium, and extra-large. The extra-large thrusters were the two OMS engines used to significantly change orbits. Generating about 6,000 pounds of thrust each, they could raise and lower the orbit significantly when burned together or singly. Each one was mounted pointing aft from their two respective pods on either side of the vertical tail. They had their own dedicated propellant tanks, but these tanks could be interconnected to burn the left engine from the right side, or vice versa. This was for redundancy as well as to give us the ability to balance the Orbiter laterally.

The medium thrusters were the Primary Reaction Control System jets (Primary RCS), usually referred to as the primaries. There were thirty-eight primary jets that produced about 800 pounds of thrust each. Distributed across the three pods—the two aft OMS pods and the forward RCS—each pod had its own tankage, and the jets were arranged on specific manifolds, numbered 1 through 4. Manifold number 5 was for the small vernier jets. There were just six vernier jets of about 24 pounds each. While the primaries were the big movers, they used a lot of gas, and trying to do fine pointing with them was bound to use lots more—so the verniers were used any time you wanted to move just a small fraction of a degree and not shake things up.

All the thrusters used what is referred to as hypergolic propellants—that is, the oxidizer (nitrogen tetroxide, a nasty orange liquid/gas) and monomethyl hydrazine (an even nastier poison)—which ignited as soon as they came into contact with each other. Ignition produced thrust in the rocket chamber. You didn't need any sort of electrical igniter—just open the valves, and whoosh—you had thrust. Of course, you wanted to make darn sure that they didn't come into contact with each other when it wasn't intended, so the plumbing systems were separate, and for our use, redundant. This made for a veritable plumber's nightmare of tubing, valves, and regulators, all of which were controllable from the cockpit switches, or through the main computers. If a jet started leaking, you generally had to close its manifold. Unfortunately, that manifold contained other jets as well—which meant that a leaker didn't cost you one jet, it cost you several.

PROP Officers spent a lot of time making sure that they knew where their fuel was, what jets were active and available, what jets had problems, and how to get the thrust needed in any direction should it be required. The PROP system, with lots of hardware and countless ways in which it could be configured, was a target-rich environment for the simulation instructors, and they took advantage of it. We often finished a simulated docking with barely enough jets to make it safely work out—and rarely any more than that.

When things got hairy in the trajectory, propulsion, and navigation worlds, it was common to see these officers huddled together, trying to figure out how to tell the Flight Director the bad news—that we could no longer accomplish the mission, and that we might not even be able to have a good way to come home. Likewise, when we knew that there were failures in the propulsion and navigation systems, and the trajectory was off nominal (not as planned), the Flight Director was able to lob the whole problem in their direction. I have to admit it was fun to watch the feeding frenzy as they worked toward a solution.

So now you have a vehicle that you can navigate and control, and hopefully have some idea where it is and where it is going. However, none of this can happen if you don't have a computer system to operate everything. Enter the Data Processing Systems (DPS) Officer and their magic boxes. It is common today for aircraft to be built using what we call fly-by-wire systems—that is, there are no direct mechanical connections between the pilot's stick or yoke and the aircraft's control surfaces.

Even engines are fly-by-wire in most machines. In this type of system, the pilot moves a control that effectively tells the computer, "I want to go over there." The computer decides the right combination of control surface movements to make that happen and carries out the pilot's wishes. It does all this with software, running in computers, and transmitted through specialized computers. It's a system that translates digital code into analog commands, or analog data into digital parameters for the computer to digest. As I said, this is pretty normal here in the twenty-first century.

But in the 1970s, when the Shuttle system was designed, fly-by-wire was in its infancy. At that time it was a monumental task to design a safe, reliable system that could operate with failed components and ensure that the appropriate functional capability would still be provided. The DPS Officer was responsible for looking over all of these computers, as well as their software, and managing the system through failure cases so as to provide the most capability and the safest redundancy.

The heart of the DPS were five identical General Purpose Computers (GPCs) that ran the software to operate and monitor the vehicle. We used three major functions: the Guidance, Navigation, and Control (GNC) software (previously mentioned), the Systems Management (SM) software that allowed us to control and monitor systems throughout the Orbiter, and the Backup Flight Software (BFS) that was used to provide a redundant software package for the critical phases of ascent and entry. No one was ever quite certain that there couldn't be a generic fault in the Primary Avionics Software (PASS) that consisted of the GNC and SM software, so early in the program it was decided that a separate entity would write the code for the BFS to make sure that a common software fault couldn't bring down all the computers. If the PASS had such a common software fault, then you could always engage the BFS, which was running all by itself in one computer. Generally speaking, the PASS GNC software ran simultaneously on four computers for ascent and entry, and the built-in SM function of the BFS was used to look at systems when required.

The GPCs of course talked in digital code, and all the temperatures, pressure, speeds, and other data generated by sensors throughout the vehicle were analog measurements. So black boxes known as MDMs (Multiplexer/Demultiplexers) were used to gather data and convert it to digital code for the GPCs to digest. When the GPCs had commands to send out, they sent the data to the MDMs and those boxes converted the data to what was needed to move a control surface or fire a jet.

The entire DPS (and the GNC system as well) was organized into four Flight Control System (FCS) channels, each with its own string of hardware that commanded redundant strings of controllers. For instance, an elevon had a four-channel hydraulic actuator, and each channel was commanded by a single channel. During ascent and entry, when the control surface was needed, each of four PCs was in control of an FCS string, and each would send its commands to the control surface actuator. So long as all four of the GPCs came up with the same commands, all was fine. If, however, one came up with a different answer, then three outvoted one, and the one channel was ignored. Flight controllers and the crew looked closely at all the channels, watching for faults—in which case they could disable the faulty channels.

Any GPC could be assigned any string—and a GPC could control multiple strings. The DPS Officer was the conductor of this "string orchestra," making sure that the crew and vehicle always had as many working channels and computers as possible. Sometimes, it was hard to figure out where a fault lay. In such cases they'd try a stringing configuration (say, put two strings on one GPC, and two others on another), and see what resulted. Sometimes the stringing could get quite entertaining—so long as you weren't the one in the hot seat. String theory is not just the latest thing in high-end physics—for decades Mission Operations hosted meetings on how to string for every foreseeable failure case. The theories changed as we learned more—just like the strings hypothesized by physicists. DPS controllers had to be sharp and keep a large matrix of options open to them during dynamic flight phases.

Once an Orbiter was in space, things got a little more relaxed. To save power, numerous GPCs were put to sleep, with generally only one GNC machine and one SM machine up and running. If we were doing something flight critical, like a rendezvous, we of course would bring more machines back up. But generally, we ran as few machines as possible, with others "freeze dried" such that they could be brought up at a moment's notice if failures occurred. Interestingly enough, there was no way to actually power down a GPC short of taking out a set of wire cutters and snipping the power lines. Even when it was switched off it was still running. And this made sense because without the computers, you had no Shuttle. It was, after all, fly-by-wire, and without the wires (the computers), it wasn't going to fly.

Closely related to the DPS was the other digital electronic discipline, with their own set of black boxes—the Instrumentation and Communications Officers (INCOs). A discipline that moved intact from Apollo, the INCOs could be thought of as the radio operators for the Shuttle flight control team. Shuttle communications were all digital, and the INCOs were responsible for voice communications (both ways, up and down), for telemetry data that came down from the Shuttle to the ground, for commands that went up to the vehicle, for television signals that came down, and for recording data on board the vehicle. There were numerous radio systems, from the plain and simple UHF aircraft radios that were used as a backup in case all the sophisticated digital gear failed, to the primary comm system that sent digital data both ways using the S-band frequencies, and up to the high data rate Ku band (another set of radio frequencies) that carried mostly scientific data and high-density television signals.

The S-band system was the primary link used to carry voice both ways. It also carried commands going up and telemetry going down. Unfortunately, the data rates available on S-band meant that not all telemetry parameters could come down at the same time, so different formats (known as Telemetry Format Loads, or TFLs) had to be chosen depending on what was going on. That meant that at certain times, some flight controllers didn't see some of their data, and selecting which format— each format contained a specific set of data parameters—became an interesting game that the Flight Director had to referee. This, for some reason, became easier as the program went on—mostly because the formats and procedure matured, and more folks became comfortable not seeing specific things at certain times. In the early days, however, there was always a lot of switching going on, with a lot of horse trading behind the scenes.

INCO's S-band antennas were fixed on the skin of the Orbiter (actually they were mounted underneath the tiles, which were transparent to radio waves). These antennas were located on the belly and on top. Each of them was fairly directional and pointed up or down, forward or aft, left or right. It was the INCO's job to make sure that we always used the antenna that was best pointed at a ground station or satellite to communicate with the Orbiter. This was computed automatically by the GNC software when everything was working. When either the GNC lost its attitude knowledge or the ability to talk with the antenna-pointing hardware, the INCO became a busy person trying to visualize the Orbiter's attitude and the relative position of the station to which we were trying to communicate. The result was that just when we needed to talk to the crew the most (to solve a problem), our ability to talk with them became the most suspect. It made for an entertaining time watching the INCO try and predict antennas when he had no knowledge of where the vehicle was actually pointing because the pointing software was being lied to by the failing software! At some point, just before losing contact, we'd just throw up our hands and tell the crew to select the "best antenna." They had a signal strength meter on board and a rotary switch that they could use to hunt for the antenna that gave them the best performance. The crewmembers were motivated to do so because until we could talk with them, they were on their own. And, of course, there were times when they liked it that way. Sometimes you just needed peace and quiet to solve a problem.

The Ku band communication system was used to transmit large amounts of data from the Shuttle, primarily to support payload operations but also to provide TV coverage. The Ku band antenna was a pointable dish that was deployed after the payload bay doors were open on orbit. It could be easily blocked by the Orbiter because it needed a direct line of sight to the relay satellites that we used to communicate with the vehicle.

For most of the Shuttle program this is the system we used. Attitude nego-tiations were always going on between the various users of the Shuttle and with Shuttle systems experts who needed to maintain communication and have particular attitudes for thermal control or micrometeoroid protec-tion. It was the Flight Director who had to referee such discussions and make the final decisions—knowing that someone was always going to go away less than happy.

INCO was also responsible for controlling and pointing the video cam-eras situated in the Orbiter payload bay. The crew had primary control, of course, but once they went to bed, and unless the cameras were dedicated for a particular payload activity, the INCO could "drive" them around to look at stuff on the Orbiter or on Earth. The original space camera that INCO controlled was the camera on the Apollo lunar rover, and the flight controller who drove it during moonwalks, Ed Fendell, became famous as Captain Video. Fendell was still doing the job in the early Shuttle days, and he handed over the duty to other generations of controllers later on. Many of the great views of Earth that people all over the world have enjoyed have come from an INCO-pointed, space-based camera.

All of those digital systems were useless without electrical power, of course. Electrical power on the Orbiter was provided by fuel cells that used liquid hydrogen and liquid oxygen to generate power. They pro-duced significant amounts of water as a by-product. Most people have heard of electrolysis, which is a process where you can produce hydro-gen and oxygen by passing electricity through water. Well, fuel cells do the same thing in reverse. The good thing about them is that the water they produce is extremely pure and can be used for drinking, cooling, or other purposes. The bad news is that the fuel cells always provided more water than was needed, so the excess had to be dumped. Once the ISS was around, the Shuttle gave that excess water to them. The ISS was a vehicle that was "water poor" because it got its electricity from solar pan-els. The panels power the ISS with free energy from the sun, but unfor-tunately they don't provide the water you get from fuel cells.

The electrical generation and distribution system was essential to every system on the vehicle. It was looked after by the console operators known as Electrical Generation and Illumination (EGIL). The EGIL was another console position that dated back to the early days of US human spaceflight, and its operators were proud of the fact that they had been around forever. They always had a bronze eagle on the console to show their pride—until, of course, some other discipline decided to one up them and kidnapped the eagle. As I recall, ransom notes started appearing for the next year or so with pictures of the blindfolded eagle statue being delivered at random. I think it eventually came back, but not before at least one picture had it deep underwater in the Neutral Buoyancy Lab pool, with its own little scuba tank.

EGIL was not only responsible for the fuel cells, they had to watch over the entire power distribution system of the Orbiter. If the DPS was the heart of the Orbiter, the electrical system was the power grid. Organized into more than a hundred individual buses, the main arteries split into smaller distribution buses then split again, and again. Each bus can be thought of like an individual circuit in your house, each with its own circuit breaker. Hundreds of circuit breakers spread through the cockpit protected the buses, along with some fuses and automatic relays. When a bus went down, it generally took multiple sets of equipment with it. As a result, every flight control discipline would be affected. It was easy to know what capability you had lost when one bus went down. It got more complicated when two went down, because now you had to analyze a two-dimensional matrix to make sure that you hadn't lost critical redundancy. When a third bus was lost, along with a flight control channel, things *really* got interesting. That is where the EGIL lived... right at the point where all of these failures combined.

Simulations were always busy for the EGIL, serving somewhat like the power company on a stormy night. With buses going up and down, and limited ability to troubleshoot (because you didn't just want to close a popped circuit breaker without a dire need—you could start a fire if a short circuit had caused it to pop), the EGIL oftentimes couldn't make anyone happy and just had to settle for the least bad of all levels of unhappiness.

The EGIL was also responsible for the cryogenic (cryo) storage tanks that held the liquid oxygen and liquid hydrogen that fed the fuel cells, as well as the heaters that warmed the gases in those tanks so that they would generate pressures to feed the cells. Back in Apollo, the cryo tanks actually had motor-driven stirring paddles that helped mix the ultracold mixture of liquid and gas so that the oxygen and hydrogen would flow from the tank. No EGIL ever forgot that it was a cryo tank stir on Apollo 13 that ignited the oxygen tank when one of those motors short-circuited and caused a spark. Using the heaters in our tanks was a similar process. Even though the lesson was learned and design steps ensured that the same thing couldn't happen with the Shuttle system, there was always a little catch of caution when turning on cryo heaters—and that caution lasted throughout the entire program.

During major power bus or fuel cell problems, the Flight Director looked at the EGIL to give him as much power as possible in as safe a manner as he could. The EGIL was also fairly deeply involved in mission direction decisions, as the cryo was one consumable that you simply couldn't live without. You never wanted to run out of cryo, or you had a dead vehicle—so managing cryo and power generation became the primary job of an EGIL Officer. They had a lot to do in the planning phase of a mission, defining who got what in terms of kilowatt hours at what times of the mission. Never tick off the EGIL—he could easily make your life miserable.

The other critical system for crew survival was, of course, life support. Now you can seal human beings into a hermetic capsule, and they will survive on the air and water you gave them at the start (this is how the earliest Russian spacecraft worked). But it is usually better to provide a way to supply oxygen, get rid of carbon dioxide, provide water (and remove humidity), and control the temperature for comfort. In short, you need to maintain a habitable environment for the crew. On the Shuttle this was done with a number of specialized life support systems managed in Mission Control Center (MCC) by the EECOM. During the Apollo program and early Shuttle days, EECOM stood for Electrical, Environmental, and Consumables Manager. But when the electrical job was split off to the EGIL, the EECOMs kept the acronym but changed the first E to stand for Emergency—for they were always in the middle of any environmental emergency situation that might develop.

Responsible for maintaining a breathable atmospheric pressurization, water supply and humidity control, and thermal control, the EECOMs operated a busy console position that watched over systems spread throughout the Orbiter. Many of their essential functions were things that the average person never even considered. While it is obvious that you need to provide air for an astronaut to breathe, it is not so obvious that one of the most essential systems on any spacecraft is the thermal control system. In short, you have a system generating electrical power, and every other system in the vehicle uses that electricity—but electrical boxes are notoriously inefficient users of power. They generally generate more waste heat with that power than they actually use. In an atmospheric vehicle, it is easy to dissipate that heat into the surrounding air. Because you don't have that air in space you have to radiate the heat away if you can, or get rid of it in other ways. Think of the entire Shuttle as a heat generator. The computers, the communication systems, the fuel cells—and the astronauts themselves—all generate heat.

To manage all that heat, the first thing you need to do is collect it. Heat collection was done using water loops and heat exchangers—you heated water by running it through plates on which a computer or another electronics box was mounted. The electronics heated up the water running through those plates, and the heated water was then carried to a heat exchanger where the waste heat was transferred to a Freon-based cooling system. This Freon was then run through large radiators mounted to the inside of the payload bay doors, which were opened once you got into space. Since black space is extremely cold, the radiators would shed the heat from the Freon, run the cold Freon back to the heat exchanger, and pick up more heat from the water—and the cycle continued.

But what did you do if the radiators weren't yet deployed or were stowed for entry . . . or if a radiator developed a leak or a blockage? The heat wasn't going away by itself, and heat was probably the biggest killer of systems if you let it get out of hand. You could manage a cabin leak by adding more oxygen or nitrogen from the tanks, but heat had to be rejected. So a secondary cooling system was developed called the Flash Evaporator (FES). The FES worked the same way that splashing water on hot skin cools you down on a warm day—the water is sprayed on a surface heated by your blood (or the hot Freon, in the case of the Orbiter) and the heat is released by the vaporized water. As a result, the working fluid (blood or Freon) cools down. In order to do this in the Shuttle, we used a supply of water and a flash evaporator surface that had heated Freon running through it. The resulting steam was vented overboard.

Obviously, this system only worked so long as you had water available—but as we said before, the Orbiter was water-rich because that is what the fuel cells generated as a by-product when making electricity. Water on the Orbiter was a commodity—you needed it to drink, use it for hygiene, make up dried food with it, use it for cooling—and it was EECOM's job to keep track of where all the water was and where it was going to go. He or she needed to make sure that there was always enough water stored for cooling the vehicle during entry after the payload bay doors had been closed. This wasn't as simple as knowing how much water you needed to cool the vehicle for the nominal timeline—they also had to plan for contingencies like going around an extra orbit or two on entry day so as to get to a different landing site—or calling off the landing and getting the doors back open to try for another day. This water management was just one of the consumables that went into the COM part of the EECOM's name.

EECOM also had to manage the atmosphere in the vehicle. They drew oxygen from the cryo tanks that EGIL was using to generate electricity—a peaceful exchange of consumables in which they got O_2 and EGIL had a place to dump their water. Nitrogen was stored in pressurized bottles in the payload bay, but rarely did the cabin require a great deal of nitrogen gas during a mission. The Shuttle was designed with an automated system to sense the percentage of O_2 in the cabin, and maintain it by adding oxygen when it dropped below a certain level. Interestingly enough, we discovered early in the program that human beings use a very predictable amount of oxygen in a given period of time, and it was just as easy to manage the percentages by opening a valve and sending a continuous "leak" of O_2 into the cabin. We could always turn on the automatic system, if we needed, or open and close valves to manage it manually in emergencies—but this controlled leak was simple and usually kept the oxygen level right where it needed to be.

Of course, as humans use O_2 they produce CO_2—a gas that becomes poisonous to our systems in sufficient quantities. So it had to be removed from the cabin continuously, using CO_2 scrubbers—canisters of lithium hydroxide (LiOH), which attract the CO_2 in the air. The canisters fill themselves up with CO_2, which couldn't be removed during a mission, so we flew many cartridges of LiOH and changed them out on a regular schedule based on the number of crewmembers and their sizes. LiOH was a consumable that had to be tracked—and EECOM had that job as well.

Humans generate an extraordinary amount of humidity as well—they breathe out moisture and exude it as sweat. This all went into the atmosphere and had to be removed on a continuous basis. This was done using a centrifugal humidity separator that spun moisture out of the air and collected it for transfer into the waste water tank. The waste water tank also collected urine for the Waste Collection System (WCS)—otherwise known as the toilet. And it was another of EECOM's responsibilities. If you're beginning to realize that EECOM was a busy person, you're right! Generally, they were at the center of any systems-level emergency or crisis—not the least of which was the case of a cabin leak, where they had to determine the leak rate and recommend ways to stop it. If they couldn't stop it, then it was their job to maintain cabin pressure using the stored gases at their disposal to get the crew home safely.

On a long mission, of course, waste water tanks filled up and needed to be dumped. In fact, the same thing often happened with the fresh water tanks before we began donating water to the *Mir* and ISS. Water dumps were conducted through special nozzles on the side of the Orbiter. It could be fascinating to watch as the liquid in the tanks sprayed out of the nozzles and instantly froze into a little snowstorm. It was fun, that is until it began to freeze to the nozzle and form ice blocks or blobs. These could plug the nozzle and/or form a debris hazard should they break off. The nozzles were heated to prevent freezing, but heaters could fail, or the amount of water dumped could overwhelm the heater—so water dumps were always treated with special care, and not a little trepidation. Once space station docking became the norm, we realized that dumping water would contaminate solar arrays and scientific experiments, so dumps became a thing of the past as a routine operation. Instead, we stored the water in special, collapsible bags. Of course, we usually had to dump some of the waste water we were holding when we got off the station—especially if we weren't going to come home right away. No one liked hooking up bags of waste water because fittings invariably leaked a few drops of the nasty stuff. But sometimes the bags just couldn't hold it all.

Propulsion, navigation computation, communication, electrical generation, and life support—these comprised most of the essential systems needed to make the Orbiter a functioning vehicle. Most—but not all. For the remainder of the critical systems, a special console position was created—the Mechanical Maintenance Arm and Crew Systems (MMACS—pronounced "Max"). The MMACS position was created after the *Challenger* accident. Before then, its responsibilities were spread out over a number of disciplines. But once this position was formed it stayed through the life of the program. Mission Control Center positions changed over the program as mission needs changed, and as we learned more about the best way to organize teams.

The Mechanical systems (the first M in MMACS) of the Orbiter were numerous and varied. Basically, anything that opened or closed with a motor was considered a mechanical system; this included the payload bay doors, the vent doors, and things like the doors that covered the star trackers for ascent and entry. The air data probes, used to sense airspeed and altitude on entry, were also swung in and out on motors. There were also nonmotorized mechanical systems—the hatches, for instance. They were manually operated—and if you ever see a movie of a Space Shuttle hatch operating with a push button, you now know that they got it wrong.

The MMACS's job also included one of the essential systems for ascent and entry—the Auxiliary Power Units (APUs) that drove the hydraulic pumps that powered the flight controls. The three APUs (one for each hydraulic system) lived in the aft compartment. Each generated about 120 horsepower from a 9-inch turbine wheel driven by decomposing hydrazine fuel over a catalyst bed. They spun very fast (about eighty thousand rpm) and were cantankerous little devils for the first ten years of the program. The hydraulic pumps supplied motive power to the flight control actuators that were necessary to fly, and you need at least two of the three of them to give full authority to fly the vehicle. Technically speaking, you still had limited control with only one running—but it was easy to run out of control if the wind conditions were anything but light. You need more control capability in gusty or strong winds, and a single APU simply didn't provide enough power in those cases.

In addition to the critical APU/hydraulic systems, the MMACS Officer was responsible for the landing and deceleration systems—the landing gear, tires, wheels, brakes, nosewheel steering, and the drag chute. All operated electromechanically and in concert with the hydraulic systems. The landing gear deployed, for instance, using pressure from any of the hydraulic systems, and those same hydraulics powered the brakes and nosewheel steering. The drag chute was developed after the *Challenger* accident. It worked in concert with the brakes to slow and control the vehicle on landing. It fell in the MMACS bailiwick, as well.

The final major mechanical system to come along and get dumped in the MMACS's lap was the Russian Docking System (APDS). This system was clearly designed by elves using magic fairy dust and built by relatives of Swiss watch makers—the number of cables, latches, cams, and gears was frightening. Yet it pretty much always worked and connected the Shuttle to the *Mir* and ISS on every flight that went to a station. The MMACS worked with the Russians to understand and operate the system, and eventually they made it a routine part of the overall operation. The system was considered androgynous—there were no male or female parts, and any APDS could dock to any other APDS. This was a great improvement over the less flexible Apollo (or Soyuz) probe and drogue: with a male and female component, and only certain things could dock with other things.

The second M in MMACS stood for Maintenance—specifically In-Flight Maintenance (IFM). IFM is when the crew is asked to fix or take something apart in flight that was never really intended to be fixed or taken apart in flight. The IFM guys were all top-notch mechanics who knew how to jury-rig systems and use tools and parts in ways that were never intended. Over the years of the program, they had the crew build things from paper, plastic, duct tape, and glue. They gave the crews procedures to jump-start computers, swap black boxes, and rework plumbing to send water where it was never intended, in an effort, for instance, to bypass a clog or a failed valve. Although the EECOM discipline owned the toilet, the IFM guys were the plumbers—and they were the ones who figured out how to fix it when it went wrong. Those IFM folks worked for the MMACS, who represented their work and products to the Flight Director.

The A in MMACS stood for Arm—they were responsible for the Remote Manipulator System (RMS—the robot arm) when it was not in use, or when it was needed in an emergency and the on-call arm specialists couldn't get into the MCC in time. That was the theory anyway. In reality, the RMS position was usually staffed any time that the arm was on board, and the vehicle was on orbit. So, although MMACS Officers were trained for the arm, they rarely did anything with it.

The final letters in MMACS stood for Crew Systems—a broad category of equipment that accounted for pretty much anything that was installed in the cabin, or was loose in the cabin, that the crew worked with. From the seats used for ascent and entry to the orange suits (and helmets, and parachutes, and survival gear...) used for the same thing, the MMACSs were the experts. They also knew how to work the galley (and how to fix it). And they had responsibility for all the loose camera gear (still and video) carried into space. They were responsible for knowing where everything was stowed, and also for the lockers themselves. The escape pole—theirs. Tools—also theirs. The Crew Systems aspect of the MMACS job pretty much meant that if something was causing trouble and no one else had it clearly in their domain—the MMACS was the one the Flight Director turned to. (If you haven't guessed already—I was the first certified MMACS Officer, and I served in that role for quite a few missions—and as the head of the MMACS section—for several years before being selected as Flight Director.) It was a great launching pad for the center seat because it literally put your fingers in everyone else's systems when it came to IFMs, and the mechanical systems that supported all the ascent and entry work.

The systems we have talked about so far constituted the core Shuttle systems—but there were always a few specialists in the room to cover specific bits of the Shuttle when required. The Booster Officer (unimaginatively called BOOSTER) was present for every ascent. The Booster Officer was responsible for the three main engines (SSMEs), the External Tank (ET), and the Solid Rocket Boosters (SRBs). The SSME was an incredible engine—generating a half-million pounds of thrust. It was throttleable and reusable—no mean feat considering that its design was conceived and baselined only about a dozen years into the history of manned spaceflight. The three engines were mounted in the back of the Orbiter, but their fuel and oxidizer came from the big orange External Tank plumbed through pipes almost 18 inches in diameter.

These pipes actually penetrated the Orbiter's belly, near the rear, and the holes had to be covered up with special doors after the ET was jettisoned. Of course, these doors belonged to the MMACS Officer, because they were mechanical and driven by motors. They were critical—if they didn't close properly, you had a hole in the heat shield, and that was always going to be fatal.

We used to joke that a Booster Officer was like a moth—they only lived for a short period of time, and when the engines shut down eight and a half minutes into flight, they were gone as quickly as they could get the engines in a safe and inert configuration, which didn't take long. Actually, they had everyone's respect because of the high energies and the rapidity with which things could go wrong in their systems. While there were many, many pages of procedures to deal with engine problems, we all felt that, in most cases, if something was going to go badly wrong back there, it was going to happen so fast that you'd never have a chance to deal with it. But we trained for what could be done, and the Boosters were always there and working to make the ascent as safe as possible.

When the Boosters went home, one of the orbit-only specialists usually moved into their vacated console space. Who took that space varied over the years. The RMS operators took over the Booster console for some years, and the Extravehicular Activity (EVA) Officer took this space later in the program. The RMS operator did a huge amount of work before every mission, planning every move of the robot arm, which was itself a complex machine with its own computers, data, and command system. The arm, built by the Canadian Space Agency, was as long as the payload bay, and it mimicked the human arm by having a shoulder, elbow, and wrist that moved much like that of a person. The end effector or "hand" was a cylindrical contraption with moving snares that could grab a special grapple fixture that was installed on a payload before flight. The universal grapple fixtures can be seen all over the ISS today as the station was largely put together while in orbit by the robot arms on the Shuttle and the ISS itself. Many years from now, the technology might well have moved on to a different type of end effector and grapple fixture—but these will remain, mementos of the Shuttle that built the ISS.

The RMS was an extremely capable—and fairly complex—piece of machinery that had significant redundancy. But it also had some frustrating failure modes that could lead you to very slow operations under some conditions. But when it was all working (which was most of the time) a skilled operator could touch an egg set in the payload bay without breaking it. Although arm motion and trajectories were planned and thought out in great detail before every flight, the controllers and astronauts who operated the arm were always coming up with new trajectories in real time when the need arose.

The arm became an essential piece of equipment in the years after the *Columbia* accident because mission rules required an inspection of the Orbiter's thermal protection system after every launch, before it proceeded to dock with the station. This was performed with cameras attached to an extension boom that was picked up by the RMS and positioned underneath the Orbiter. The Flight Day 2 Inspection became routine, and the arm proved its worth mission after mission until the end of the program.

The final system specialist (and special system) that was essential to orbit operations was EVA—the guys with the space suits. Space walks have been a part of the US manned spaceflight program since Ed White floated out of the Gemini capsule for America's first twenty-minute walk in space back in the 1960s. Learning to work outside the spacecraft was essential to getting to the moon, and it is amazing to think that this technology, which was developed in such a short period of time, would allow humans to walk on the moon and then ride in a buggy on the moon. After Apollo, because we were no longer going to be on a lunar surface, space suits became specialized for use floating in a zero-G environment. The Shuttle usually carried two suits until it was time to build the ISS—then the great "wall of EVAs" appeared. They were used on a seemingly endless series of space walks that were required to build the station, assembling it from components brought up in the payload bay. That wall seemed daunting to those of us who lived through the early years of learning how astronauts could work in space—but darned if it wasn't conquered in style with all the ISS assembly EVAs completed successfully.

Once the ISS assembly got underway it was not uncommon to have four space walkers on a given Shuttle mission, so up to four suits were put aboard. We carried two in their usual spots in the airlock, and two more were stored in the Mid-deck. Sometimes, crewmembers were close enough in size that they could share a suit. If so, they only had to swap out the soft components (the lower torso, legs, arms, and certain internal pads) between work sessions.

The EVA space suit or Extravehicular Mobility Unit (EMU) was itself a complete little spaceship, comprising a full life support system, instrumentation, communications—in later years it even had an emergency propulsion system that could be used if an astronaut found themself suddenly free-floating, untethered in space. Such an event was unlikely since astronauts were trained from day one in tether protocol (always being attached with two tethers). But the fact that the ISS could not maneuver to pick up a separated astronaut forced the development of the SAFER—a little emergency jet pack that was attached to all EMUs during the ISS time frame.

The suits really were marvels of engineering, with pumps, oxygen systems, cooling, and communications all rolled into one package the size of a human being. The torso was rigid, and you entered by breaking the suit at the waist. The lower portion—below the beltline—was soft. You got into the lower portion before snaking your arms and head through the upper torso portion. It took care (and a long checklist) to get into a suit and be ready to go outside—nothing like the movies where a person hops into a suit, zips it up, and finds themselves floating in a matter of minutes. It just doesn't work that way because there is really no margin for error.

The EVA Officers in MCC are the same people who build the procedures for each mission, and they train the crew to perform the specific tasks needed for each flight. They are truly an integrated, all-around operations engineer—involved with the design of the equipment, developing the timeline and procedures, training the crew, and then watching over them when they are actually in space. EVAs are dangerous because you have a person outside with nothing but a nonredundant suit to protect them—but the dangers are mitigated because the EVA Officers have been involved with the specific mission plan, the hardware, and the crew from the beginning. They truly are experts in that specific day's work—and it shows.

The many and various systems officers who knew the Shuttle intimately were, of course, only part of the team. Equally important were the guidance and navigation flight controllers, the trajectory officers who knew the mathematics and physics of spaceflight like the backs of their hands.

Chapter 2

Spaceflight 101

Orbital Mechanics

The phrase *orbital mechanics*—like *nuclear particle physics* or *the theory of relativity*—is something that makes many people's eyes glaze over. The average person thinks that subjects like these are far too complex to understand. But if you strip away all the math and simply try to understand what is going on, they are actually not that hard to grasp. As for orbital mechanics, it is the physics of spaceflight—it describes how objects move relative to one another in space. Once you understand the mechanics of how planets and spacecraft interact with gravity, it's easy to make a spacecraft do what you want it to do.

If you take a baseball and throw it straight away from yourself, it will eventually hit the dirt—it runs out of speed, and then gravity pulls it to the earth. If you're a major league pitcher, your throw will probably go farther than mine because it launches off your hand faster. But gravity eventually does its thing, and the ball will come down to the earth. Now, let's build a machine to throw the ball faster—it will go farther, of course...but the ball will always curve down and hit the dirt. That curve the ball follows, by the way, is known as a ballistic path. Because of gravity, any object thrown or fired into the air will eventually be pulled back to the earth. With every increase in speed, the ball goes farther. If we change the ball to a projectile, and start firing it from a gun, then it will go farther still.

Naval guns can fire projectiles so far that they go over the horizon before falling back to the earth (hopefully on their targets). That phrase, *over the horizon,* is key to understanding orbital mechanics. The spherical shape of Earth means that as you travel horizontally, the surface is always curving down. If you can throw the object fast enough that its drop, its ballistic path, is equivalent to Earth's curvature, then it will never hit the planet—and voilà—it has gone into orbit!

It takes a tremendous amount of speed to reach that point where your object is not going to hit the earth—in the neighborhood of 25,000 feet per second (that's over 17,000 miles per hour) if you're talking about flying in the region known as low-Earth orbit (LEO). Let's call that anywhere from about 100 miles above the surface to about 400 miles—give or take. That's where the Space Shuttle did all its work. If we take our imaginary baseball or projectile and keep upping the speed, eventually it will go so fast that it never falls back to Earth. The speed at which that occurs is known as escape velocity. We never worried about escape velocity with the Shuttle—it didn't have the capability to go that fast. Unlike the Apollo spacecraft that took men to the moon, the entire Shuttle spacecraft was always coming back to Earth. It would have taken approximately its own weight in fuel to propel the Orbiter to the moon. (I asked my navigators to figure that out one night...)

So the first element of orbital mechanics isn't that hard—make an object move fast enough relative to the planet, and it will never hit the planet. It will just keep going around and around and around—until, of course, something slows it down to where it begins dropping toward the surface. What might slow it down? Well, one thing might be running into the thinnest wisp of the atmosphere, the widely spaced molecules of gas that reach for hundreds of miles into space. The atmosphere doesn't just end abruptly, it gradually gets thinner and thinner as you move away from the planet's surface. It never really goes away completely, it just eventually fades in density until it no longer has any effect. Just about 100 miles above Earth's surface, there are enough air molecules that if a spacecraft runs into them, an infinitesimal amount of energy is lost with every collision. You can't measure the energy from any one collision, but if you add up enough of them you can eventually discover that you have lost some speed. And that speed loss adds up.

The slower you go, the more you fall back to Earth. So the effect of running into the atmosphere is that it drags you back down. If you don't add some velocity with a rocket motor burn every once in a while, you won't stay in orbit. It doesn't take a lot—just a couple of feet per second every day—but if you don't account for it, your mission isn't going to last very long. We use this effect to our advantage, of course—it's how we bring a spacecraft home. If you point your spacecraft so that when you ignite an engine it slows the craft down significantly, then you will drop lower into the atmosphere where the gas is thicker, which then slows you down even more—eventually to the point where you have been captured by the atmosphere, and you fall to the earth. This is what happens when a meteor becomes a meteorite: it generates enough heat from the friction of the atmosphere to quickly burn up, which creates the streak we see in the night sky. Of course, burning up on reentry is the last thing we want a spacecraft to do, so we have to enter in a controlled fashion. We'll talk about that later.

For now, let's remember that if we go fast enough horizontally, we end up orbiting Earth rather than falling back to the surface. The faster you go, the higher you go. The slower you go, the lower you go (until you fall out of orbit and are captured by the planet's atmosphere). These are the basics of orbital motion (or mechanics). With that mental picture, you can understand almost anything else we'll talk about when it comes to the Space Shuttle's trajectory.

To get the Shuttle into orbit, you have to do two things: get it out of the atmosphere and accelerate it to orbital velocity. In the simplest terms, the first part of that is done with the Solid Rocket Boosters (SRBs). These two monsters had enough energy to loft the entire Shuttle stack up to an altitude where the air was so thin as to be negligible. They did this in just a little over two minutes. Each SRB put out about 3 million pounds of thrust, for a combined total of 6 million pounds. By comparison, the three Shuttle main engines contributed another half-million pounds of thrust each, for a total of 1.5 million pounds—a much smaller portion of the overall thrust available at liftoff. Not insignificant, of course, but still—the SRBs dominated during what we referred to as the first stage.

When the SRBs dropped off, the vehicle was up where the air was not a real factor, but it was only going a couple thousand feet per second horizontally. It was now the job of the main engines to accelerate the ever-lightening stack (the Orbiter and External Tank) to that magic number of 25,000 feet per second to get it to stay in orbit. That "ever lightening" part is important—as you burn fuel, you get lighter, but with a constant thrust (about 1.5 million pounds, remember?) you are going to accelerate more quickly. That's a consequence of Newton's basic law of motion—if the force remains the same and the mass decreases, the acceleration goes up! Now the Orbiter was designed for a maximum acceleration of three times the force of gravity (3 Gs). A "G" is about 32 feet per second per second, so 3 Gs is just about 100 feet per second per second—you're really gaining velocity quick! When the acceleration reached that point, the only thing you could do to keep from over stressing the vehicle was to throttle the engines back, and that is what we did. In the last two minutes or so, the throttles would come back to make sure we didn't break anything.

In the simplest terms, when you reached the desired velocity to make orbit, you shut down the engines and coasted into orbit. It sounds simple—but it isn't. Let's stretch our knowledge of orbital mechanics a bit. Let's assume that you are in a circular orbit—the same altitude above Earth at all points in the orbit. If you decide that you want to go higher, then you have to increase your speed. This is done with a thrust event, known as "doing a burn," because we thrust by firing—burning—an engine. If you squint and allow yourself to approximate, doing a burn to increase your speed by 1 foot per second will increase your altitude by about a half mile. That's not an instantaneous gain—what you are actually doing is driving yourself uphill until you reach that new altitude, which you will reach when you are halfway around Earth. But you won't remain there. Think of the ballistic path that a ball takes when you throw it—it first increases in height, then gravity pulls it back down, and so it comes back down again. The same thing happens when you increase the Orbiter's velocity—it will go uphill to the new altitude, but it eventually comes back to where it started... right to the altitude where you increased the speed. It so happens that you'll reach your new altitude halfway around the world, and then you'll be back where you started when you complete the orbit. It will continue in this elliptical path for as long as you let it.

However, if we want to raise the orbit all the way around, we can simply thrust again by the same amount when we reach our new height (referred to as the apogee). We will have raised our altitude at the starting point by a half mile as well, meaning that we will be in a new circular orbit a half-mile higher than when we started the pair of burns. We will have also increased our velocity by 2 feet per second total.

The math is really convenient if you're trying to do it in your head— if you want to raise the orbit by 10 miles, you simply burn 20 feet per second (fps) initially, then another 20 fps at the new apogee, and voilà— you're in a new circular orbit 10 miles higher, and it cost you a total velocity change (referred to as delta V) of 40 fps. Raising and lowering the orbit is how you execute a rendezvous. But for now, it's important that we get into, and then know how to get out of, orbit.

Let's take a look at the very end of the initial launch. There we are, thrusting all three main engines, accelerating at 100 feet per second every second. Knowing what we now know about orbital mechanics, we know that for every second we burn the main engines at this point, we are raising the orbital altitude by 50 miles when we get around the planet. We need about 100 miles of altitude to reliably stay out of the atmosphere—consider it the minimum safe altitude we want to end up in. The International Space Station is at an altitude of about 200 miles, the Hubble Space Telescope is about 350. The Orbiter lived in the altitude band between 100 miles and 350 miles—a difference of just 250 miles. In terms of orbital insertion speed, that is just five seconds of burn time.

A full ascent, from the launch pad to Main Engine Cutoff (MECO) was about eight and a half minutes, or about 510 seconds. The orbital altitude range of the vehicle meant that cutoff would be plus or minus five seconds, which is a very small percentage of that total burn time. Miss it by 1 percent and we were either not in orbit or we were going way too high, without enough thrusting capability to circularize the orbit—or to get home. So MECO was critical—you had to time it exactly right in order to get precisely into the orbit you wanted, and we considered precise to be within a couple of miles.

No problem, right? I mean, throw the switch at the right time to cut off the engines, and you've got it made. Well it's not that simple, because you don't just shut off an engine that's putting out a half million pounds of thrust. It doesn't go from full thrust to zero in an instant—it tapers off. Every engine tapers off a little differently, so you need to know the exact shutdown characteristics of each engine. You can measure these characteristics on a test stand on the earth, and then use that information to figure out when to command the engines off, in fractions of a second, so that you end up with exactly the amount of thrust you need to end up at the target altitude. When you look at the hundreds of variables involved, you quickly realize that it's going to take a lot of smart people to figure this out. Fortunately, there are lots of engineers and physics guys who *are* smart enough to model it and come up with the right answer. In the earliest of rocket flights, way back to Mercury, they were happy just to know that they'd made it into orbit. In the Apollo and Shuttle eras, we needed to have precise control of where we were going to end up—and the accumulation of rocket flight experience made that possible.

Now those who are following closely have already figured out that spaceflight is much more complicated than this. Recall the basics of orbital mechanics, remembering that what goes up must come down. If we have thrusted ourselves from the ground up to an orbital altitude, say 200 miles, we are only at our apogee. Like throwing a ball straight up into the air, we're going to be coming back down to our starting altitude. This will happen when we get all the way around Earth. The nitty gritty, of course, is that we need to do some shaping of the trajectory to make sure we aren't going to come all the way back down to the ground.

Remember we need to travel horizontally at a high enough speed so that we don't fall back to the surface. If we really want to end up in a circular orbit (and not an ellipse), we need to do a burn about halfway around the planet. Since we go around the planet in ninety minutes, it means that forty-five minutes after launch we need to do that burn—and we can't use the main engines to do it. For that, we switch to our Orbital Maneuvering System (OMS) engines.

The OMS engines are mounted in pods on the upper rear corners of the Orbiter. There are two of them, and each produces about 8,000 pounds of thrust. You can burn them together, or separately, depending on how much thrust you need and how finely you want to manage the final velocity. These engines burn propellant stored in their pods, the same kind of fuel and oxidizer used for the attitude control jets—in fact, the tanks for the OMS and Reaction Control System (RCS) jets can be shared (or interconnected) between the two systems, if need be. Once you have shut down the main engines, and jettisoned the big External Tank (about eight and a half minutes into the flight), the OMS and RCS are all that you have. Because these engines are so much smaller than the main engines in terms of thrust, you have to burn them longer to get the same amount of velocity change—in fact, the acceleration available from the OMS engines is barely noticeable inside the vehicle.

But getting back to circularizing the orbit. In the simplest terms, when you get halfway around Earth from your launch site (over the Indian Ocean when launching from Kennedy Space Center), you point yourself in a direction to thrust ahead, and then burn the OMS engines to add the velocity that you need. It can be in the neighborhood of 100 feet per second or more. In the earliest Shuttle flights, we were happy to see that it worked to keep us in orbit—later on, we had learned enough about trajectory shaping and burn times that we were pretty unhappy if we missed our orbital parameters by more than a couple miles. We used ground and space tracking to confirm the orbit we needed to be up in, and we added what we learned into planning for every flight until we became very good at putting the Orbiter exactly where we wanted it to be.

Once you made it to orbit, changing that orbit was simply a matter of adding or subtracting velocity by adding speed with a burn or taking it away (you did that by thrusting backward). When you wanted to come home, you needed to thrust backward. That's called retrograde. This is where we got the term *retrofire*, which was done in the early years of space travel with retrorockets. The backward thrust lasts until you slow down the vehicle enough to lower its perigee to about 80 miles, which is where you are assured of being captured by the atmosphere. If you decrease velocity by too much, you lower the perigee by too much, and that means you enter the atmosphere too steeply. Too steep of an entry means that you decelerate too quickly and have to dissipate your orbital energy in a shorter amount of time, which means that you get much higher temperatures on the skin and you burn up. Too little velocity change means that you enter the atmosphere at too shallow of an angle, and you could effectively skip off its surface, much as a rock skips off a pond. The problem with a skip is that you scrub off speed, which makes you slower, so you drop back into the atmosphere more steeply the second time, and eventually you end up with that steep entry again and burn up—sort of how those stones always drop into the water and sink.

By now, however, you can see that to bring the Orbiter home from a 200-mile orbit, you need to drop the perigee (the low point of the orbit) by about 120 miles (200 minus 80), and to do that, you need to slow it down by about 240 feet per second. If you filled the Orbiter's tanks before launch, then you had approximately 600 feet per second total orbital maneuvering capability (the total delta V) that could be used throughout the mission—to raise and to lower the orbit. The key to mission planning and execution was to use that delta V wisely.

Spacecraft Evolution

The earliest manned spacecraft were, by today's standards, pretty crude. Both the Russian Vostok and the American Mercury were essentially pressure vessels that could keep a human alive as the craft was lobbed into a low orbit, circled around Earth a few times, and withstood the heat of reentry to come down safely to the earth. The craft had limited maneuvering capability. Mostly they had the capability to orient themselves, or control attitude, so as to point windows and cameras in particular directions. They had the capability to execute a reentry (we now call it a deorbit) burn using retrorockets dedicated to the purpose. They were simple but effective vessels that carried out their missions.

Spacecraft technology matured and evolved relatively quickly. The Gemini series of US spacecraft included not only the ability to control attitude, but to change the altitude of the spacecraft using maneuvering engines. With this capability came the ability to rendezvous with other spacecraft or targets—but in order to do this, they needed to be able to navigate—so sensors and computers had to be added to allow the astronauts a way to figure out where they were, where they needed to go, and how to effect the necessary changes to close the gap. This same maneuvering capability provided them a wider and better control of where and how they came back to Earth, which made piloting an essential part of spaceflight.

Of course, if you were going to stay in orbit long enough to do anything useful, you needed more than just a sealed can of air to do so. You needed life support systems to keep the air fresh and circulating, and you had to control humidity and temperature. And to do these things, you needed power—which meant not only an electrical system to distribute power but also the ability to generate that power, be it from storage batteries, fuel cells, or solar arrays. Of course, if you are going to keep a human alive for a few days in space, you need to feed them and provide them with water, and have to provide ways for them to eliminate waste. You also need the humans to be able to communicate with each other, and with the earth. From those requirements follow all sorts of mechanical and electronic systems that can keep the humans occupied, doing useful work, and recording data so that the missions have value in the long term.

After the Gemini program came Apollo, and with it came an increased complexity by an order of magnitude. You now had two crewed spacecraft—one to take the humans to the moon, and another to get them down to the surface. You had space suits that allowed the astronauts to exit their craft and work outside (these were much more sophisticated than the ones that allowed the astronauts to do this from Gemini, where they were attached to the spacecraft for essential life support). You had multiple rocket engines that had to work to get the men down to the moon and back up. You had different navigation schemes for getting to the moon, landing there, and leaving. Just the technology of the science experiments themselves represented entirely new levels of complexity to operations.

The Space Shuttle was conceived in the time frame of Apollo and was intended to be part of a multipronged system to allow routine access to space. Originally, the Shuttle was to be built concurrently with a space station and a space tug—the station was to be a place to take humans, and the tug was going to be used to move satellites and other spacecraft from Shuttle orbits to other orbits where they could do their work or launch to other planets. In order to serve a set of yet-to-be-defined missions and objectives, the Shuttle became a very complex vehicle, far more sophisticated than anything that had come before, and far more capable as a result. In order to serve the widest possible array of missions, the Shuttle had to have the widest possible array of capabilities—and that meant a wide variety of systems.

Aerodynamics

Early spacecraft designers realized that they had two problems from a physics and aerodynamics standpoint. The first was how you accelerated an object from zero to orbital velocity. They solved this with the rocket—a way of channeling the high energy of what amounts to a controlled explosion into a propulsive force that will last long enough to put you in orbit. The second problem was, however, somewhat harder—that is, decelerating the object and returning it safely to the earth. The laws of physics say that if you are going to put all that energy into an object, you need to figure a way to get it back out of that object—without melting or otherwise destroying it in the process. Meteors are a good example of something that fails to make it through the atmosphere safely—they enter the atmosphere at a high speed and usually at an angle that creates so much friction that they burn up completely before making it to the ground. This is fine if you're trying to prevent falling rocks from hitting people on the head, but it's a bad thing if you want to bring a spacecraft home in one piece.

The first generation of crewed spacecraft took lessons from the earlier designs of nuclear-tipped Intercontinental Ballistic Missiles (ICBMs). Nuclear devices are complex devices that don't react well to extremely high temperatures (until, of course, they generate those temperatures themselves). Launching a hydrogen bomb on a missile and lobbing it halfway around the world meant that you had to get it back into the atmosphere without it burning up—so that it could do its thing on the intended target. The earliest rocket work by the Pilotless Aircraft Research Division (PARD) at NACA (National Advisory Committee for Aeronautics, NASA's predecessor) Langley in the late 1940s and into the 1950s was dedicated partly to solving the problems of reentry heating. This led directly to better ICBM nose cones, but also to the shape used to design the first Mercury spacecraft: a blunt body with an ablative coating that burned off with high heat, carrying the heat energy away so that the soft, pink human body inside made it all the way to the ground. Well, not to the ground but to the water, actually; or more accurately yet, to the point where a parachute could be deployed to lower the craft into the ocean.

Similarly, the Russians used blunt bodies or spheres to reenter. They ended their flights with parachutes as well, although the very first manned missions required the cosmonaut to eject from the capsule and use a personal parachute for the final landing.

If you reentered using a blunt body with a symmetrical design, and maybe you even introduced a little roll about the line of symmetry, you had a predictable reentry trajectory over which you had little control. You knew where you were coming down, but you couldn't change it. It was soon discovered that if you had an asymmetrical design, you actually could generate some lift—and you could point the lift vector in different directions to effectively change your impact point. You now could control the vehicle's final touchdown spot (at least by a little bit). Gemini used such an asymmetry by offsetting the center of gravity, and so did the Apollo spacecraft. But even though you now had some control over your landing point, it was limited—and the parachute system still required a water landing to keep the final impact loads survivable for the crew. The Russians land on solid ground because they fire a final braking rocket at the moment of touchdown to cushion those last loads—but it is still a pretty good impact.

The dream of almost everyone in the modern spaceflight era (most of us had been inspired by science fiction writers in the mid-twentieth century) was to have a fully reusable spacecraft that could land like an airplane, on a runway, and be flown again. It was recognized that the use of parachutes and water landings required a tremendous amount of recovery resources and was impractical in the long run if you wanted routine access to space. In order to effect a routine landing capability without parachutes, you needed not only the capability to move your target point around, you also needed to generate enough lift to break the descent rate—to flare for a landing, in other words. But wings are not great at dissipating the heat of reentry, so having them in space didn't mean you were going to have them after the fiery part of reentry. What was needed was a stubby wing that could survive the heating and be around to execute the flare for landing.

NASA had worked with wingless, lifting body reentry vehicles in the 1960s. Although they worked, they had very high landing speeds and unforgiving control characteristics. So work began on developing a truly winged vehicle that was shaped to survive reentry and provide a reasonable amount of lift so as to be controllable and provide a reasonably soft touchdown so that the vehicle could be reused. The Shuttle Orbiter was the result of that research.

The Shuttle has been described as a heavy glider. It is, in fact, a glider for entry and landing, generating lift with its wings as it reaches the thicker part of the atmosphere. Gliding capability is often referred to as a ratio of lift to drag. Whereas a typical light airplane might have a lift-to-drag ratio of 10 to 1, and a sailplane might be in the neighborhood of 25 to 1 (or far better—in the case of competition sailplanes, 50 to 1 is normal), the Shuttle has a relatively poor glide performance that topped out at about 4 to 1. In truth, it usually flew in the 2 to 1 regime. That means, in aeronautical terms, the Shuttle was a rock. The joke about a low lift-to-drag airplane is that if the engine quits and you're looking for where to land, just throw a brick out and follow it down. The Shuttle was a lot like that brick, but when you compared it to a ballistic reentry, it was a heck of a good glider.

Now there are a few challenges with launching a winged glider into space and bringing it back. First, there are potential aerodynamic load problems on ascent. In order to make the Shuttle light enough to fly significant payloads into space, it had to have a fairly light structure. The airframe was therefore not a fighter plane that could be thrown around the sky with abandon—you had to be careful not to exceed the design limits—and that included aerodynamic loading. Almost everyone has stuck their arm out a car window and felt the lift and drag generated by cupping your hand like a little airfoil. The faster the car goes, the more work you have to do to keep your hand from blowing backward. The same thing is, of course, true with airplanes—and the Shuttle was an airplane. The problem was that if you were going to travel at speeds approaching or exceeding Mach 1 (the speed of sound), you could easily rip the wings off with the air loads.

That's exactly what happened in the *Challenger* accident—the leaking solid rocket motor cut a hole in the bottom of the External Tank, which exploded. The explosion threw the *Challenger* off the fireball at a high angle of attack (the angle between the wing and the oncoming air) and that put so much pressure on the wings that they failed—followed, of course, by the rest of the structure. I have often been asked if the crew could punch off the External Tank in a situation like that, and the answer is always the same—take a look at the *Challenger* crash and you'll see the results. The aerodynamics of winged rockets on ascent is very unforgiving. In the Shuttle, we had very specific limits to angle of attack and sideslip (the angle of wind hitting the nose from the left or right) and very little margin of error. This is why upper air winds were an important consideration on launch day—a wind shear (a sudden change in wind speed or direction over a short altitude span) could easily create an overload situation in the vehicle.

Entry, of course, is where people really think about aerodynamics. The science and engineering that went into the design of the Shuttle's entry trajectory was a tremendous challenge and was the result of the accumulated knowledge of thousands of aerodynamicists, thermodynamicists, and every other "cists" you could think of. Once your rocket engines had done their thing to drop you out of orbit, you had to prepare to kiss the atmosphere at Mach 25—25,000 feet per second (or about 18,000 miles per hour). At that speed, hitting gas molecules rips those molecules apart into a plasma—a super-heated gas. It is like applying a blow torch to the skin of the vehicle. Since the Shuttle was basically built like an airplane, there was no way the bare structure could survive such treatment, so the thermal protection system, a layer of silica tiles and reinforced carbon, was applied to protect the primary structure. These tiles were amazing—applied with a special adhesive, they could withstand the intense heat of entry without transmitting the heat to the aluminum to which they were attached. Unfortunately, these tiles and carbon were also relatively fragile, and they had to be treated with care—as we found out throughout the program, and tragically with the *Columbia*. The bottom line is that without the thermal protection, the Shuttle would have been vaporized in short order upon an attempted reentry. Most people who have an active interest in the Shuttle program have heard that before.

But there was more to entry aerodynamics than that. It was important that you hit the atmosphere at the proper angle. Too steep and it would capture you quickly, overloading the vehicle, breaking off the wings, and you would burn up and/or break up. Too shallow and you would literally skip off the atmosphere, enter again later, but this time too steeply... with the same results. The appropriate angle for the Shuttle entry was right about 40 degrees nose up, and you had to hold this through the hot part of the entry profile. Control was accomplished with maneuvering jets, and they did a good job of holding the necessary attitude. At the same time, the computers were recognizing increased loads on the movable control surfaces and had to adjust these surfaces to keep the vehicle in the appropriate attitude.

Now if all you did was hold the appropriate angle of attack and went straight ahead, your impact point with the earth—if you did nothing else—would be essentially determined. The beauty of a winged vehicle with lift capability is that you could point that lift in various directions to let you turn left or right. As you did this, you could also vary the direction of the lift vector to make your impact (or landing) shorter or longer. In fact, the Orbiter's landing footprint from the time you did the deorbit burn was variable. It allowed us to land up to 750 miles on either side of the nominal aim point, 500 miles short, or about a thousand miles long. This footprint was sort of an odd-shaped ellipse that gave us plenty of capability in choosing landing sites.

The lift vector was pointed in various directions by rolling the vehicle left or right. If you rolled it into a steep left bank, you decreased the component of the lift vector that generated an upward force, and so you sank more quickly—increasing the rate of descent. You still went somewhat straight down the entry line, but you got a little bit of a left turn out of it. Eventually, you turned enough so that you were starting to point away from your intended landing site. When that happened, it was time for a "roll reversal"—you rolled from a left bank into a right bank, and that started a gentle turn back toward the runway—and eventually beyond it. Then you'd do another roll reversal to head back where you wanted to go again. By doing these rolls, you could vary the descent rate and manage energy and heating, all at the same time. It was a clever way to maintain a constant angle of attack yet still control where you were headed.

As the Orbiter dropped lower and lower into the atmosphere, it slowed down, and at the same time it got more bite from the thickening air. Wings respond not to airspeed as much as they do to what's called dynamic pressure. Dynamic pressure is the product of air density multiplied by the speed of the aircraft squared. It is directly measured, just like it is when you had your hand out the car window—you can feel the resulting force. The more dynamic pressure you have, the more force you have acting on your wings and control surfaces. These forces provide you with the ability to control the aircraft. But you have to be careful with generalization like that, because odd things happen when you are flying at supersonic speeds. Shock waves form on various parts of the aircraft, and they can reduce the effectiveness of controls by effectively shadowing them from the airflow. This is sometimes referred to as the controls being "blanked out." Sometimes, the controls will actually reverse their functions at certain speeds. This makes for real problems for pilots—it can be hard to figure out just how to move the controls depending on what speed you're going—and the workload would be over the top. Fortunately, in the age of computers, we have the ability to put a machine in between the pilot and the controls. The pilot says, "I want to go left," and tells the computer this by moving the stick left. The computer looks at the current speed and dynamic pressure (and a bunch of other factors), and realizes that, at this point in time, the best way to go left might be by moving the controls in a counter-intuitive direction—but because all the aerodynamics have been written into the computer's program, the net result is a left turn. The pilot has no idea what is actually happening with the controls and doesn't really care.

The Shuttle flew entry pretty much like that—with a combination of maneuvering jets, the control surfaces on the wings, and the rudder and speedbrake on the tail, all being controlled by the computer, based on the pilot's (or autopilot's) wishes. In fact, most all entries were flown by the computer (autopilot) until the speed dropped to the subsonic regime, for reasons that will be discussed later on. The important things to take away from all this is that bringing a winged vehicle back from orbit is not a trivial task, and it took the combined knowledge of generations of aeronautical engineers to pull it off. It was a magnificent accomplishment that is often overlooked by those without a lot of background in supersonic flight.

In Mission Control, we had several positions devoted to navigating and guiding the Shuttle. Just like in the Apollo days, these folks lived in the front row, which became known as the Trench. The name is a holdover from the original control rooms, where this was the lowest row in the room, down on the floor. Even when we built the new Control Center with a flat floor, this front row was still referred to as the Trench—some traditions never die (and they shouldn't).

The Trench was anchored by the Flight Dynamics Officer (FDO—pronounced "Fido"). Their job was basically navigation—figuring out where the vehicle was and where it was going. They were responsible for computing burn targets, they knew how to get us to landing sites, and they were the ones who kept track of everything having to do with orbital parameters, trajectory timelines, landing sites—you name it. If you wanted to know the best time to launch—ask FDO. If you wanted to know the current temperature at a landing strip overseas—ask FDO (the weather guys worked for him, as did the Landing Support Officers). If you wanted to know if you're going to be in sunlight or darkness for a particular time on orbit, or if the vehicle is speeding up or slowing down—FDO is your person.

The FDO has always had one of the largest back rooms in the Control Center because they have lots of things to do and keep track of. When it came time to distribute mission crew patches (a traditional gift from the astronauts on a mission to the flight controllers and Mission Control Center personnel) to a team, most disciplines asked for a couple patches, maybe three, to cover themselves and their back room. FDO asked for about twenty or twenty-five—they sometimes seemed to have their entire office supporting every mission. That's not a problem, actually: we wanted everyone in the operation's organization to participate in missions. It made everyone aware that what they did contributed directly to our prime goal.

FDO lives in a world of vectors, and everything in this world is expressed in terms of vectors. We know where the vehicle is in space because we can express its position as a vector—a description, a particular time, where the vehicle is (in three dimensions), and where it is going (also in three dimensions). If you know these elements (time, position, and velocity), and you know the equations of motion due to the laws of gravity, then you can use math to add the motion to the original position to tell where the vehicle is at any point later than the vector's time. FDO can then compute the effects of future burns on the position and predict where the vehicle is going to be at some future time. FDOs are good at math—in fact, many of them aren't engineers, but mathematicians and physicists.

FDOs who worked the ascent and entry shifts had additional responsibilities: determining abort parameters for ascent and watching over the entire entry trajectory during landing. The FDO position contributed a number of their alumni to the Flight Director office, as it was a position that encompassed a large portion of what went on in Space Shuttle operations.

Sitting right next to the FDOs were the Guidance Officers (GUIDO—pronounced "guide-o," not like an Italian hit man). They were responsible for understanding and monitoring onboard guidance systems and determining how well they were doing. They also had all the responsibility for procedures used by the crew in navigating. It was often hard to separate responsibilities between the FDO and GUIDO positions, which is one reason why they sat next to one another. As a Flight Director, you could just lob a question in their general direction and the right person would pipe up with an answer.

GUIDO was used primarily for ascent and entry in the Shuttle program, and was replaced by the Rendezvous Officer (RNDZ) for orbit operations that included rendezvousing with another object or spacecraft. As an Orbit Flight Director, I rarely worked with an Ascent/ Entry GUIDO, but I spent much of my career talking to RNDZ flight controllers. They watched over the onboard targeting and maneuvers, made sure all the i's were dotted and t's were crossed in the Rendezvous Checklist, and were experts in the esoterica of flying two spacecraft in close proximity. It's nowhere near as simple as driving close to another car on the freeway—orbital mechanics gets very interesting. When you try to meet up with something else going 18,000 miles per hour around a planet, having help from a team of experts like these is essential to making it work.

The Third Leg of the Triangle

We liked to divide Shuttle mission control teams into three broad categories—Systems, Trajectory, and Operations. The one we haven't yet talked about is Operations. Operations included all the folks who were dedicated to planning and executing timelines and serving the particular objective of the mission—the payload. Although everyone on the flight control team (including the trajectory and systems folks) served the mission, the operations team really had the reins when it came to making sure that the purpose of the flight was achieved. That group included the Flight Activities Officer (FAO), the Payloads Officer (referred to simply as Payloads), the Flight Director (Flight), and the Spacecraft Communicator (CAPCOM). There were others who came and went, and we'll cover them as required—but this group had the overall job of beating the drum to make sure that everyone was pulling at the same tempo, and in the same direction.

The FAO's primary product was the timeline—called the Flight Plan—which was the detailed listing of activities that each crewmember (and each control center) had to execute in order to make things happen. Sometimes, things went exactly according to that plan. These plans included voluminous information on which procedures had to be executed at what times, as well as communication plans, trajectory operations (such as maneuvers), and just about anything else that was important to execute at a particular time. It even included items that weren't time dependent but had to be executed each day—things like cleaning air filters or reading the mail. We learned to give the crew this list and let them pick up these tasks whenever it was convenient for them.

The funny thing about the timeline was that it either worked perfectly or it was a complete disaster. If all the experiments and payload tasks worked, then you could execute the timeline from start to finish and make few changes. But if an experiment had problems or if an operation took longer than planned or if some other monkey wrench got thrown into the flight, well, all bets were off. It was the job of the FAO (and their busy backroom support folks) to replan the mission, while making sure that all the various dependencies (what needed to be done and in what order, and which things had to occur at particular times in the orbit) were met. The Shuttle timeline was a complicated beast, often a true house of cards that could only be put together one particular way. Well, actually, there were multiple ways to put it together so that it would work—but there were many, many more ways that it wouldn't.

Early on in the program, the crew launched with several copies of the thick, overall detailed timeline. It was good for them to have this, but when the timeline had to be replanned, that got thrown aside. Eventually we learned that the crew really only needed an overview for the entire mission, so they launched with that—we sent up detailed pages every morning based on the results and replanning from previous days. Later in the program, with the advent of better data and communication techniques, we simply sent up files as PDFs that the crew could use on their laptops. These files could be printed out as they desired. Regardless of the method, the FAOs were always in charge of the timeline. They were like the editorial staff for the morning newspaper.

The FAOs owned no systems on board the Shuttle, but theirs was an important job. One of their backroom positions was Pointing— determining the Shuttle's attitude at any time during the flight. Once a spacecraft is in orbit, it matters not (aerodynamically) which way the nose is pointed. But payloads needed to be pointed at targets, stars had to be put in the field of views of star trackers, and certain parts of the Orbiter needed to see sun or shadow depending on thermal requirements. It was the job of the Pointers to make sure that all the inputs that went into how the Shuttle was oriented were met—or at least could satisfy as many conflicting requirements as possible. Their attitude timeline went into the morning flight plan along with everything else, and the pilots keyed in maneuvers as required.

Now the Payloads Officer position was one that morphed and changed a bit from the beginning of the Shuttle program through to the end. Originally, it was simply a NASA flight controller, whose job was to represent the payload customer in the MCC's front room. Of course, that sounds simple, and conceptually it is, but it was a huge job because the Shuttle rarely (if ever) simply carried one payload on a flight. Generally, there were one, two, or a few major payloads on every mission—but there were also scads of little things that went along for the ride. Sure, you were carrying the Hubble Space Telescope into orbit, and that was what everyone saw. But there were always little Mid-deck experiments that had to be dealt with—from a vial of fish eggs to sometimes baffling medical experiments—and each one had to have a knowledgeable presence in the MCC so that when there were questions, comments, or problems, we could get them dealt with and make the most of the time for the payload.

I always reminded my flight controllers when they were pooh-poohing a minor experiment that it probably represented some scientist's life work. These scientists had probably begun looking into a problem in graduate school, proposed an experiment that would answer the questions they had, and then spent a dozen years developing the experiment and getting a grant to put it on the Shuttle. They would then spend ten more years analyzing the data they got and writing about it. That's a significant portion of a professional life and was not to be dismissed lightly. So everyone got the full support of my flight control teams, and the Payloads Officer was their advocate in the front room.

Later on in the program, commercial satellite deployments went away, and the emphasis became less on commercial endeavors and more on visiting and assembling space stations. We still carried payloads, but they were really components that had to be put together as part of an International Space Station (ISS) construction project. In addition to these major components, we had lots and lots of small cargo—supplies for the growing station, experiments that were going to be worked with by the *Mir* and ISS occupants, and all the other bits and pieces you needed to support a crew in space. In addition to carrying cargo, the new components that were being added to the space station had to have knowledgeable representation in the front room in regard to assembly and checkout.

So the Payloads position morphed and became the Assembly and Checkout Officer (ACO). The ACO was the single point of contact within the Shuttle Control Center for all questions relating to what we were carrying in the payload bay or in the cabin for transfer to the ISS. They stood with one foot in the ISS world and the other in the Shuttle world and bridged the gaps between the two. It was as monumental a job as it sounds. The ISS quickly became a monster in terms of size and complexity. The ACO and their back room worked closely with the ISS team and the Shuttle FAOs to make sure that all the assembly tasks were completed on time, and that cargo was transferred in both directions according to the plan. You didn't want the Shuttle to come home with cargo that was intended to be left on the ISS—especially if it was some important small part that was desperately needed on orbit.

In addition to the FAO and the various incarnations of Payloads Officers, there were several other positions that worked in the Shuttle MCC that got little notice outside the operations team. Most of them resided in the back row—which didn't mean that they weren't important.

We usually kept the Flight Surgeons back in the corner, not because we didn't appreciate or like them, but because in that spot it was hard for anyone to look over their shoulder and see what was considered to be privileged medical data—the vital signs on the astronauts. It was no joke that these were matters of legal privacy, but of course if a medical event came to have an effect on the mission, the Flight Surgeon would share that information with the Flight Director, oftentimes over the airwaves, so as to keep such things off the communications loops. It was the Flight Surgeons' job to know and understand each astronaut and advise the Flight Director if something was going on that might affect the mission. They could do this just by listening to the crewmember, but a better way to really get a feeling for each crewmember's health was through a private medical conference, or PMC.

PMCs were the exception to the rule that everything going on between the ground and the spacecraft was generally available to the public. NASA did not operate Space Shuttles in secret (except, of course, when the mission was being flown for the Department of Defense), and we tried very hard to keep everything decisional out in the open. Reporters who weren't familiar with us often thought we were hiding something—I think they had simply seen too many science fiction movies—but for the most part it was easier to keep things in the open than it was to try and keep them hidden.

To conduct a PMC, the Flight Surgeon and a few selected folks (selected by the Surgeon) would set up in a conference room and the air-to-ground loops were routed to that room—and nowhere else. PMCs were scheduled every few days, or they could be called by the crew. A Flight Director could participate, but had to do so from the conference room. Occasionally, mission impacts might be discussed, and when this happened the Flight Surgeon would fill in the Flight Director. Crews were generally very good about not trying to slip things past the team by using a PMC, and the various commanders made sure of that. Commanders and Flight Directors understood the responsibility of keeping everyone in the loop when it came to mission impacts—nothing good could come out of people marching to different tunes.

The Flight Surgeon was also a welcome addition to any team because if you came down with a headache or indigestion, you couldn't just call a halt to your work and go home. For this reason, the flight control team was under the Flight Surgeon's charge as well. They kept a medical box handy—the Surgeon's Little Box O' Pills—that gave the staff what they needed to keep controllers on console when they were needed. Generally speaking, most of the stuff in the medical box was over the counter, and it was certainly appreciated when some back room's spaghetti feast started to have adverse effects.

In the other corner of the back row, we kept the person who most of the public considered to be the MCC—or at least the voice of Mission Control: the Public Affairs Officer (PAO). A PAO was always on console when we were flying a mission. They didn't support many simulations, but they were the ones whose voices went out on the mission commentary on NASA TV. Whether they were giving a play by play of the ascent or entry, or occasionally telling folks what was going on during a slow overnight shift, it was the PAO who had to interpret what was going on for the general public. The PAO could choose what video went out— either from the spacecraft cameras or from the cameras in the MCC— and tried to keep everyone out of trouble. They also were a frequent point of contact from the outside world to the flight control team, even if they didn't want to be.

I remember one night on STS-95, the flight that carried John Glenn on his second space mission (the first since his Mercury flight), when my PAO walked over and got my attention. It seemed that they had been forwarded an email by the headquarters PAO, who had gotten it from the NASA administrator, who had gotten it from a White House staffer— who had been asked to send it by President Clinton. The message was that the president wanted to send a personal greeting, privately, to John, congratulating him on making it back into space. He didn't necessarily want the message to be public—it wasn't about publicity—just a note to a friend.

The first thing I did when I read the note was, of course, to wonder if it was real. So I tasked the PAO to go back and use another string of people to verify that the email addresses were valid, that the people in the chain were real—and that they were truly connected to the White House. This they did, and sure enough—the message was authentic. So we dropped it into the email uplink to John. These were the kind of things that fell into the PAO's playbook under the heading of "other duties as assigned."

It was a broad book, and they were often quite busy trying to keep everyone inside and outside the control center happy and informed.

Of course, I would be remiss if I didn't also acknowledge the many hundreds of support people who made the MCC itself work. Ground Control (GC), led by an officer who was located in the front corner of the room, owned everything in the MCC building and out into the global communications network. If the data or commands weren't getting to or from the vehicle—we called GC. If someone's console was on fire, that was GC's problem. And if the vending machines were empty on a Sunday afternoon, well, you could complain to GC. But the truth was he didn't have much pull with the food vendors. However, I knew GCs who cared enough about their responsibility to the team that they would send one of their backroom people to a nearby grocery store to pick up what controllers needed if necessary.

From the computer and communications technicians to the building maintenance workers, and all the way to the janitorial staff—the MCC didn't function unless *everyone* functioned. In fact, I made it a point that when mission patches were handed out to my teams, no one got a patch until Olga—our long-time cleaning lady—got hers first. It was only fair, since she was there when we came to work to launch a mission, and she was there when we went home after landing. No one was unimportant— Mission Control was a team effort.

Chapter 3

Being a Flight Controller

The average flight controller for the Shuttle program started with either NASA or one of the Shuttle contractors right out of college. If they were with NASA, they might have started out as a co-op student before they left school, because for years that was the only reliable way to get an actual government job with the agency. Regardless—anyone starting out as a flight controller started out young, with only a few exceptions. Flight Operations were all-consuming and required a deep time commitment—not just eight hours a day for five days a week. New flight controllers were generally people who stayed late to study, lived in a one-bedroom apartment, and had little else going on outside work, at least until they reached the point where they were certified to work console.

I started with NASA as a co-op student in 1980, before the first Shuttle flight. I was one of a few hired at the beginning of the new wave of staff expansion to support the Shuttle program. The Johnson Space Center (JSC) had downsized considerably at the end of the Apollo moon program, with many talented people being cast aside through reductions in force (RIFs). This formed what was called the "saddle curve," with lots of employees in their mid- to late-fifties and a lot of young folks—but few in the middle. I was fortunate to get in at the beginning of the new buildup because it put me ahead of a large number of young engineers who joined the program in the early 1980s.

Like a few others, I found myself placed in this fascinating organization known as Flight Operations. It didn't take me long to realize that these were the folks who ran space missions from Mission Control, and that the individuals I was working with had directly served in the Control Center for Apollo and Skylab. Many were veterans from before Apollo, having come aboard when NASA was first learning to fly in space with Mercury and Gemini. I learned the culture and thought processes of flight control by sharing offices with men who had controlled the moon flights and who had kept humans in space for months on Skylab. I heard the stories of how they essentially created the space program from nothing—making it all up as they went along, making mistakes while figuring out better ways to do things. It was an amazing group of mentors for a kid not yet out of college.

Flight control, I learned, was about knowing more about the system or discipline for which you were responsible than the people who built it. Oh, there were engineers who understood the details better than you might, and they could be called on for fine-toothed-comb troubleshooting when needed. But most of those design engineers didn't see the big picture—they didn't understand how their equipment might be used in space. They often thought in terms of what was good for their system or component, but not how it melded with all the other systems to achieve the bigger goal of a spaceflight. That integrated regime was where flight controllers had to live. It wasn't just your system or your box that was important—it was accomplishing the overall mission that mattered.

Like most engineering students, I thought of engineering in terms of discrete disciplines. I was an aeronautical engineer, for instance—I studied aircraft design and operations. Others studied mechanical engineering, and yet others studied electrical or chemical engineering. But in the operational world, we couldn't afford to be so specialized. In order to be successful within Flight Operations, you had to understand not only your own system but how it interacted with every other system on the spacecraft. More than that, you had to understand your system's place in the timeline as well as its relative importance in a crisis. You had to learn when it was your time to lead the way and when it was time to follow. And you had to learn to think very, very fast.

Spacecraft in low-Earth orbit move at about 5 miles each second, and there is no way that you can simply pull over and park it while you try to figure things out. Once you leave Earth, the clock is always ticking. The first rule of thumb was that you always had to have a plan before you lifted off—and if everything went the way you expected, then you executed that plan, and took joy in the fact that it all went well. You met your goals—life was good. But the important part about having a flight controller's mind was that in an off-nominal situation you could evaluate a wide number of options in the blink of an eye and not only be able to head off in the correct direction but be able to explain why it was the best of all the options—or at least one of a family of options that would take the mission in the correct direction.

Potential flight controllers either had or developed this quick mind—or they simply didn't make it. Some likened it to the ability to play chess, a game where you need to look ahead a number of potential moves and try to figure out which set of moves is most likely to achieve an advantage over your opponent. But in chess, you don't have the same time pressure that you do when you're hurtling through space. Sure, a timer is involved in chess— but you have minutes, not seconds, to evaluate the possible moves. In the MCC (Mission Control Center), there was nothing more frustrating than having a slow thinker in a critical path between you and what you needed to achieve. It happened though, and the team usually suffered because of it. These folks were eventually weeded out, given office assignments, or moved to jobs where slow, methodical thinking was important.

Flight controllers also had to be able to learn and absorb new data quickly and be able to apply what they learned in new ways without a lot of shelf life. In training simulations, failures were the name of the game—that is, we had to learn how to manage numerous failures. Any failure could easily interact with many different systems within the vehicle. Power and data bus failures were the quickest way for the training team to throw a grenade in the room. Suddenly, your redundant systems didn't have the same redundancy that they had before. Where you might have had three ways to accomplish a necessary function, you might now have only two—or one if things got really bad. In effect, you no longer had the same vehicle operating under the same procedures and rules as you had when you lifted off. And you couldn't lament that fact, wish it away, or hope for better—you had to work with what you had left and move forward.

Young people learned quicker and had a larger capacity for new learning than those who were older—that was simply a fact we all understood. We piled workbooks and systems drawings onto our new flight controllers until they could recite the connections between their systems and others in their sleep. We then dumped books full of nominal and off-nominal procedures all over them and expected them to know the critical ones by memory so that in a time-critical situation they didn't have to look them up. The typical flight controller had several book shelves of information stuffed into their brains. Most of them also created their own cheat books—we called them "goodie books"—filled with their own interpretations of the many volumes that they were responsible to know. These notebooks were what they used on console. Although management hated to admit it, it was no use trying to use someone else's goodie book—each was written in someone's particular mental language and organized by the way they thought (there was no configuration control on these things). There was intense competition in flight control, but the intent was not to beat someone else—it was to be the best that you could be. That being said, you were eventually ranked against others, so the competition for the best flight assignments and advancement was real.

An Early Start

I reported for duty at the Johnson Space Center in late September 1980. With three years of college behind me, I would have been starting my senior year if I was still in classes—but I had elected to take a Cooperative Education Student position with NASA, and that was going to add a year to my college career. As I would find out later, it was probably the best thing that ever happened to me. Even though I was going to end up spending thirty-four years in a hot, sticky, Texas climate (which I did not enjoy), I was also going to get the ride of my life—and do things I never thought possible.

I was on track at the University of Minnesota to get my degree in aeronautical engineering. My plan had always been to seek a job somewhere close to home designing and building airplanes. Alexandria, Minnesota, was home to the Bellanca Aircraft Corporation, a place I had visited when I was a teenager. I was helping restore J-3 Cubs at an airport near my home as part of an Explorer Scouting program, and we took a couple of field trips to visit Bellanca during those years. It was a chance not only to see how real airplanes were built, but a chance to get there in airplanes; the airport was a short hour's flight from the northern suburbs of the Twin Cities.

Bellanca was amazing because of the craftsmanship that went into the wood-wing, steel tube, and fabric Super Viking. This was a place I wanted to work. I had no idea if they would ever be hiring, and I really didn't know what an engineering job would be like in a place like that, but it was old-school aviation and I wanted to be a part of it.

Then, just as my junior year of college was coming to a close, I read the terrible news that Bellanca was declaring bankruptcy. I was crushed—this meant that I might have to decide to live in Wichita! No offense to those from Kansas, but Wichita (home of Cessna, Beechcraft, Boeing, and an assortment of aviation supply companies) was not exactly my dream destination. It's a long way from the lakes and forests of my youth and it just didn't seem to have the kind of activities I loved—skiing, camping, and water sports. I never really considered going to work for one of the large aerospace firms. I was a small-airplane guy. I wanted to be a part of everything—design, engineering, construction, and, hopefully, flying.

What I didn't realize until later in life was that bankruptcy was part of the normal life cycle of small airplane companies. They succeeded, failed, and were resurrected on a regular cycle. Bellanca did, in fact, continue building airplanes on and off for many years after I thought they were gone. But hey—if I'd waited for them, I'd have never gotten involved with NASA, so it's hard to complain.

Fortunately, another opportunity presented itself, one from a totally unexpected direction. As I chose my usual seat in the front corner of Professor Stolaric's class on supersonic flow, I listened to him talk about a sheet of paper he had in his hand. "This announcement came to me from the College of Liberal Arts," he said. "NASA is looking for some sort of interns. If anyone is interested, I'll leave this here on the table." He said this with not a little disdain—I am not sure if it was because the liberal arts college had sent it over, or he wasn't that happy with NASA, or he just felt that anything learned outside the classroom couldn't be all that valuable. Our department was heavy on theoreticians and very light on actual experienced engineers.

In that moment, my life changed. I think that I knew it at the time. I had this sudden realization: "I should look into that!" I was a person who had a plan, a path. But suddenly, I was ready to step off that path, take a right turn, and see where it took me. I didn't even know what part of NASA was looking for students, or even where these internships might be located. But I remember the idea taking over me like I had turned on a switch.

As soon as class was over, I was out of my chair, and up to the table—and I took the piece of paper so that no one else could grab it. I don't know if I had any competition, but I wasn't going to let someone else have a crack at this opportunity before me. A thick set of government forms took precedence over homework the next week, followed by a trip to the local police station for fingerprinting. Then it all went into a big mailer, and off it went to NASA.

When the envelope came from Houston, I was shocked to see that I had been accepted, and that I was given reporting instructions for September, coinciding with the start of our fall quarter. Suddenly, instead of being a live-at-home student, I was going to have to move away, find housing, and live on my own. It was a sudden and drastic change—but something inside felt that it was right, and worth it. I showed up in

Houston on one of the hottest days ever recorded there, riding my 400cc Yamaha, with a backpack strapped to the seat. I had carefully packed a small box of items before I left Minnesota so that my parents could ship it to me after I found a place to live. I had ridden for a week on a little tour—going out to Salt Lake City to visit my sister, then heading south to the Grand Canyon before angling back across Arizona, New Mexico, and the vast expanse of west Texas. When I hit Houston's Katy Freeway, my first thought was that all those people in cars were trying to kill me— this really was life in the big city!

I reported that first day along with a bunch of other co-op students and new employees. We spent the day going through paperwork, selecting insurance options, getting badge photos taken, and receiving a general orientation to working for the government. In the afternoon, we were finally taken out to meet our new supervisors and see where we'd be working—and what we'd be working on. Up until that point, I had no earthly idea what I'd be doing; I just hoped it would be interesting.

It turned out that the way the co-op program worked was that applications were passed around to the various organizations that had the need for a little extra help (that wouldn't count against their limited staffing numbers), and supervisors flagged the folks who looked like they might be a good fit. I remember sitting with my first supervisor, Dick Koos (I later found out he had been one of the simulation supervisors back during the Apollo missions), and we talked a little about my background. He was interested in my diving and flying much more than my grades. This was probably good, because my GPA at that point was not all that impressive. What had impressed him, apparently, was that I had real-world experience in operations—taking equipment into the field, the water, and the air, and dealing with what happens in real time to alter your plans to accomplish a mission.

The group I was assigned to, the one Dick supervised, was the Payload Computer Section of the Payload Operations Division. The Shuttle had not yet flown, but the Flight Operations Directorate was already organized to not only test-fly the Shuttle but to prepare for the coming era of repeated access to space for scientific and commercial purposes. One of the major ways in which the Shuttle was to support these efforts was with the complement of Spacelab hardware—a modular package of systems designed to provide a shirt-sleeve environment in the Shuttle payload bay. The payload operations team would also support all the various experiments with power, cooling, data processing, and other needs.

Our section oversaw all the data processing equipment—the Spacelab Command and Data Management System, or CDMS. If it was doing digital stuff in the Spacelab, it was ours. There were two major computer systems—identical except in function. One supported the Spacelab subsystems, and the other supported the experiments. They were operated in parallel and they shared resources, like display units and interfaces to the command system. Using the command system, the ground could send up instructions to operate the various systems in the Spacelab so that the crew could concentrate on their science. The heart of both of the computer systems was the Input/Output (I/O) unit for each one. While they were internally redundant, there was only one of each, and they were essential to making the Spacelab work. If either the Subsystem I/O unit or the Experiment I/O unit went up in smoke, then half of the CDMS was out of business, and you were done.

There was a computer for each I/O unit, plus a backup computer that could be swung either way. The computers were simple machines with, as I recall, a Motorola 6502 CPU and 128K of RAM. Interestingly enough, the processor was the same as what was used in the Atari 400/800 consumer-grade computer you could buy in the stores in the early 1980s. Back then, the IBM PC had not yet come out, and you had your choice of Atari, Commodore, or the mighty Apple II—the really pricey option that a co-op student like me wasn't going to be able to buy. In fact, as a co-op student, I wasn't going to buy anything more expensive than a pocket calculator. It wasn't until I came on full time that I was able to buy my first Atari—and I did that only so I could teach myself machine language programming. Well, that and play *Missile Command*, as I recall... But back to the CDMS—there were a number of subsystems I had to learn, but the real thing they wanted me to work on was the Spacelab Instrument Pointing System (IPS).

The IPS was a payload-bay-mounted, gimballed platform to which you could mount astronomical telescopes and science packages that needed to be pointed very precisely. Think of this as a precursor to the Hubble Space Telescope, which was in development but was years off from launching. The idea behind the IPS was that you could mount telescopes to it, carry them up in the Shuttle, do a week or two of science, and then bring them back. The next time up, the IPS could carry improved versions of those experiments, or something entirely new. Because the Shuttle itself was a fairly crude pointing platform, with stability measured in degrees and rather abrupt jet firings that jiggled and jilted the structure, the IPS acted sort of like a Steadicam—sensing its position in inertial space and isolating the telescopes and experiment packages from the Shuttle's crude motions.

The heart of the IPS was a gyroscope package, which did the basic stabilization. There was also an accelerometer package mounted near the base that sensed motions before they got to the payload, so the system could effectively damp those motions out. There was also a package of three-star trackers that looked out at the universe to find known navigation stars and determine exactly where (in inertial space) the payload was pointing. This combination of sensors provided data to the software that computed necessary commands to the motorized gimbals to keep the whole thing pointed where you wanted—with an accuracy and stability of approximately an arc-second. The entire package was built in Europe—as was the entire Spacelab. NASA had written the original specifications for what the IPS (and the Spacelab) should be able to do, and how it should work—but then it was up to the European Space Agency to design and build the thing. Then it all came back over to our side of the pond, and we had to figure out how to actually make it work for us in space.

This was actually a good working example of what Flight Operations—and flight control in general—was all about. Design engineers built amazing hardware and software to accomplish a mission, but it was up to us to figure out how to make it do what we wanted, and how it could fail. The first step was to do system-level drawings that could help us understand its functionality, as well as provide reference materials for troubleshooting in simulations and flights. A system-level drawing is not really a schematic—it isn't designed to show actual physical wires—but it is far more complex than a block diagram.

System drawings show how power, data, and commands flow through a black box, or a whole system. These flows give the flight controllers the information they need to watch over the function of the system and figure out where failures might be occurring when things aren't working as they should. These system-level drawings are done by flight controllers, for flight controllers—and also for the crew and the folks who will train both the crew and the flight controllers. Using the original schematics as a basis, along with requirements documents and long discussions with the original builders, these drawings generally become the basis for all in-flight communication about failures and how systems work. I always said that real value in these drawings was the work that went into producing them.

On my first co-op tour, I was assigned a desk in an office with a couple of Apollo veterans—James Bodmer and Glenn Cress—who were now working on the IPS. Both were experienced flight controllers, and I think they had both been trained as electrical engineers. There I was, a traditional aeronautical engineering student—wings, tails, control surfaces, and engines being my interest and forte—being asked to produce a drawing of an electronic accelerometer package! This package featured solid-state accelerometers and the signal conditioning equipment and power supply circuitry necessary to generate acceleration data and feed it to the computer to use in the overall control loop. Now I wasn't completely uneducated when it came to electronics—after all, I did have my electronics merit badge from my scouting days, and I had built a few things using transistors, resistors, and capacitors. But here I was suddenly immersed in the world of digital electronics and Boolean algebra—something I had never had a chance to study. I think my introduction (taught by Bodmer) took about thirty minutes. Then I was marched over to a neighboring office where another engineer had a collection of Texas Instruments reference books for all the available chips that were being used around the world at that time. They had each chip listed by number, a drawing, circuit information, and—most importantly—a truth table that showed what you'd get on the output side for any combination of inputs. These weren't programmable microprocessors with a wide variety of possible outputs—these were simply logic gates that made circuits work.

So my job was to study the available schematics of the accelerometer package, along with all the notes and design requirements. I'd then look up the various chips on the schematics and figure out how they worked. I'd then figure out how they were being used in the circuits, and what they were doing—then distill that information down into a box on the drawing that showed what inputs led to what outputs. It was a satisfying game of cat and mouse, figuring out exactly what the designer had intended and how things worked. It was all logic and tracing signals—and it taught me what I really needed to know to understand how systems worked. Before my first six-month tour was up, I had figured out the IPS accelerometer package was a warm-up to tackling the Optical Sensor Package (OSP)—a set of three-star trackers that were pretty complicated (for a rookie). The really fun thing about the OSP was that it was built by Sodern, a French company, and all the drawings we had were in French, a language with which I only had a passing familiarity from working with French diving gear. But technical language is full of English cognates, and once I figured out a few key words, the rest started to make sense.

It was a fun winter, and I learned a lot. But the best part was that, every once in a while, I was able to stick my head up and realize just where I was—I was getting ready to work in Mission Control! As a matter of fact, our division's offices were on the third floor of Building 4, and the other half of the floor was the Astronaut Office. I could hardly believe that I was sharing the stairs, elevators, and restrooms with our astronaut corps—a heady experience for someone who had been watching the space program since early childhood. I never really thought that NASA would be something I could be a part of. But now here I was, rubbing shoulders with the men and women who were going to fly on the Space Shuttle, and a couple who had walked on the moon.

It was during this first tour of duty that I was shown the training library—essentially a small storeroom filled with shelves and shelves of documents on the second floor of our building. It was here that I fell in love with the idea that I could really learn a whole lot more as a co-op student than as a college student in class back at the University of Minnesota. Here I found a treasure trove of material that I wanted to learn. There was an entire series of workbooks on orbital mechanics. Another series was dedicated to teaching the ins and outs of telemetry and command systems. There was a whole stack of training manuals on how to operate MCC consoles and other hardware and software tools I'd use as a flight controller. And then there were the huge number of workbooks on Shuttle systems—including the massive DPS HW/SW 2101—everything that could be pulled together about the computer systems that ran the Shuttle. It was a couple inches thick, with closely spaced text and lots of system-level drawings.

When I was first introduced to the library, I picked up a few books that quickly thrilled me, and I soon began filling my extra hours—including lunchtimes and evenings—reading those books. By the end of the six months, I had worked my way through almost all the Shuttle systems, all the MCC training manuals, as well as books on the basics of spaceflight. I probably stuffed more new knowledge into my head during that half year than in any other time of my life—but it was all stuff that I wanted to learn, and it came easily because of that.

Mission Control was really a computer system masquerading as a large building. Essentially three stories tall (very *tall* stories), the first floor was where all the computers and communications systems lived. The second and third floors were control centers—a Mission Operations Control Room (MOCR) in the middle, and smaller support rooms all around. The floors and ceilings were massive raceways for wiring. When you went inside, you really were in a computing complex. By today's standards, that complex was exceedingly primitive—but in its day, it was truly state of the art. Most of the technology that the flight controller interfaced with was left over from Apollo. The consoles were really dumb terminals in a way—they displayed data on light panels and black and white TV monitors. The controllers could interface with the Mission Operations Computer (MOC) on the first floor with educated buttons that called for specific functions to occur.

Changing console functionality really meant writing a set of specific definition requirements and sending them to the organization that configured and ran the building. After a long approval process, someone pretty much had to go out with a soldering iron and make new connections from the consoles to the MOC. Specific consoles had unique buttons that could tell the computer to initiate particular pieces of computer code. Nothing was really reconfigurable in real time. You could select a function for a specific event light panel by dialing in a code that told the MOC what program to run. The MOC would then activate a certain set of lights on the panel. Then you had to pull an envelope of films out of the console drawer, select the film for the configuration you had selected on the MOC, and slide it into a holder in front of the lights. If you got the wrong film, the lights you'd see would make no sense at all.

Some of the control buttons on each console were standard—a panel to select TV channels and display numbers, for instance, was basic. Every controller had one. You could use it by dialing in a number—one digit at a time—and pushing a button to tell the MOC that you wanted to look at a specific display, or you could peer in on a specific display channel that someone else had already selected. At one time, there were eighty channels available in the building, and each console position had a certain number allocated to them.

So a particular controller might bring up their main display, and the MOC would put it on an empty TV channel. Anyone in the building could now look at that display by bringing up the channel. The console that originally selected the display had control of that channel, however—so if they flipped to something else, anyone watching that channel would see whatever new display was brought up. As the space vehicles got more complex, the number of parameters to be monitored went up. That meant that the total number of displays went up. But the number of available channels remained the same—making for full channels. This would force the Flight Director to ask, "All flight controllers, drop to your allocated channels!" You see, each discipline was allowed the exclusive use of a certain number of channels for any flight phase, but they were on the honor system to stay within that allocation. What do they say about honor among thieves? Well, it was pretty common to "accidentally" go over your allocation, especially when you got busy solving problems. Then someone would complain, and Flight would "request" that everyone check how many they were using. If that didn't create space, the channel police would start calling every discipline one by one until they got everyone back in bounds. Of course, that took time, and if you just needed to check that additional display for a few more seconds… you were unlikely to get caught. It was a game, and some played it pretty well.

The MOC was sort of a virtual thing—there were five computers located on the first floor, any one of which might be the MOC on a specific day or simulation (sim). The other computers might be supporting other flight activities, a sim, or development work—and they got used as required. In the simplest terms, the data stream from the vehicle was fed into the MOC in a serial stream—all the bits in a single line. Imagine, for instance, a bunch of eight-bit words, all lined up in a row. Each word represents a measurement. As the data stream came into the MOC, it was "chopped up" like a bunch of sausages coming off an assembly line, and each measurement was stuffed into its own little mailbox. The various programs running on the MOC could then say, "Hey, I need the cabin pressure measurement from sensor number one," and they would go to that mailbox and get the latest value of the measurement and use it for its computation cycle. The mailboxes were refreshed at various rates. Essentially, there was a 128KB stream coming into the building for regular Orbiter telemetry, and some parameters were updated every second, some at broader intervals, and some more frequently, such that they updated multiple times in each second.

The telemetry formats could be easily changed to support different flight phases. For instance, most of the ascent flight control data were unnecessary once the vehicle got on orbit, so you switched to a format that supported orbit operations. The key was that when the Orbiter changed format, the MOC had to follow—basically to make sure each side was using the right "decoder ring" so to speak. It took quite an army of people to keep the whole thing running and in sync, which was one of the many things that changed as the Control Center was modernized and modularized. The once-busy back rooms of the ground controllers (the folks who made the building work) became empty rooms of quiet computers, with nary a tech to be found. But that was much later in the program—in the early days, when I was a fresh-faced controller, the ground controllers making it all work probably outnumbered the flight controllers!

I did two total tours as a co-op student and accepted a job in the Flight Operations Division as a junior flight controller upon graduation. I arrived in Houston the day after the fourth Shuttle mission landed and President Reagan declared the Space Transportation System fully operational. It was a stretch to call any flying machine that had only flown four times fully operational but, hey, we just put our noses to the grindstones and made missions work. I picked up where I had left off training to be a backroom controller for the first Spacelab mission, where I was responsible for the Subsystem Computer. I spent countless hours in the Control Center training for the job. I spent many more hours in the office building learning procedures and flight rules, drawing and studying schematics, and learning how everything worked together with other systems and the timeline. For the first time, I was introduced to spending hours in the simulators where we tested our procedures to see if they worked in both nominal and off-nominal situations and configurations. And then I got to spend an entire mission working to understand how the Spacelab CDMS *really* worked—along with everyone else who thought they had understood it before launch. Nothing in the real world works quite the way it does in simulations, you see—the only way to truly figure out how the real thing operates is to use it in space.

With Spacelab-1 successfully behind us, I was called into the office and told that I had flown my last mission as a backroom flight controller—because I was needed in the front room for the next one! I was shocked but obviously smiling inside—all the extra studying and work had paid off, and I was going to get to play in the big leagues sooner than I expected. What surprised me at the time—but what I understood later on—is that this was not uncommon. When I looked around the room on my first front room simulation, I saw others of my "class" who had also studied hard and been moved up quickly. It was clear that hard work got you noticed and doing well got you promoted. I knew then that the folks who had joined the organization at the same time I did and were putting in the time to excel were the ones who were going to be moving up with me on the way to the center seat—the Flight Director chair.

My first mission in the front room was also the first mission of the Instrument Pointing System. I had built my thick goodie book, and I had flown so many simulations that it was hard to keep track of them all. I was considered the expert on the system in MCC—to the point that when the crew had significant questions, they asked for me to be put on the console with the CAPCOM (Spacecraft Communicator) so that I could help them understand all the procedure changes we had developed during the mission to overcome faults in the software that were only found after launch. As I recall, I was at home watching the mission on the NASA TV channel when I heard them ask for me by name—and suddenly the phone rang. It was the on-console Flight Director, himself, telling me I had fifteen minutes to be in the front room—and I'd better have a tie on!

That mission was exciting, exhilarating, and terrifying all at once. The system had its problems for numerous reasons relating to how it was developed and our inability to test it "all up" on the ground, but we were able to overcome them with hard work and numerous software patches that we uplinked to the computer. It went from being a basket case to a working pointing system in just a few days, and the team used it to conduct good science for the rest of the mission. I looked forward to our next IPS flight when we'd have a much better system right from the start. We trained for the flight—a solar pointing mission—for months and we were next in line, in the on-deck position, when I stopped by the lobby to watch the Shuttle launch that was immediately before ours. It was the *Challenger* on mission STS-51L. When I saw the forked-tail plumes of rocket exhaust as the solid rocket boosters flew out of the expanding cloud of gas that had been the Orbiter and our friends, I knew we weren't going to be flying anything again for a while. Having been involved in aviation since my teens, I had experienced the loss of friends in flying machines before, but it still gave me that awful, empty feeling of helplessness that can leave you feeling lost. For many of my young coworkers who didn't grow up in dangerous activities, this was a shocking, sudden lesson in the reality that what we did could cause real death. I could see that on the faces of others, and I could feel the shock inside myself. You deal with it, but you never get over it.

I was at a point in my career when my management and I recognized that it was time for me to broaden my experience once again, so it was during the two and a half years of downtime after *Challenger* that I changed disciplines and moved over to the Mechanical Systems Section. I became deeply involved with the improvements to the landing and deceleration system—the tires, wheels, brake, nosewheel steering, and the development of the new drag chute. It was my kind of thing! This was airplane stuff—the stuff I had been born to do. While I assumed that I would be the junior flight controller in the section (and that is how I started), before I knew it I had been promoted to the front room and was flying the key Ascent/Entry shift on the first flight after the *Challenger* accident. It was again an exhilarating and terrifying position, but learning to suppress fear was one of the important elements of flight operations—and I was getting good at it. Fear doesn't help you make good decisions; it just gets in the way. You learn to work through it, think of your options, and do what you can to support keeping the vehicle intact, the crew alive—and see the mission through to its end, if you can.

Training to be a backroom flight controller is all about learning the systems, procedures, rules, and interfaces to a level of detail that boggles most people's minds. We expected a good system controller to know every telemetry bit, the meaning of every digital word in a command string, and what size wire was used in the wiring harness of a black box. They needed to know exactly how it operated and how it was tested; they had to know where the official limits were and where the unofficial limits were—and just how much extra performance was available beyond that. And they needed to know, technically, how it all worked with any other system that it touched.

A front room flight controller had to know all of that—but more importantly, they had to serve the overall mission with the equipment for which they were responsible. While the back room could concentrate on what was best for their system, the front room had to know what was best for the mission—and sometimes that meant sacrificing their own equipment or system functions for the greater good. The only way they could make informed decisions was to know as much about the overall vehicle as they could. Those folks who kept looking inward at their own systems when they transitioned from a support position to the front room simply didn't last very long.

So what, exactly, does a flight controller do? Well, system controllers (those responsible for specific Shuttle systems) watch data to make sure that there are no faults and that procedures being executed by the crew are done correctly and on time. They keep tabs on all their interfaces and they contribute to the planning process, for instance, by checking procedural references in flight plan revisions. Those systems controllers with consumables (like fuel, oxygen, water, and so forth) are always keeping tabs on usage to make sure that something isn't going away faster than planned and that there will always be enough of everything to complete the mission. In the case of failures, the systems controllers are responsible for troubleshooting, following the crew through procedures when appropriate, and creating new procedures when needed.

Systems flight controllers are engineers for a reason. A person who knows how to execute canned procedures is a technician, and technicians are valuable contributors to a team. But when a problem arises and a technician reaches the end of a procedure and hasn't solved the problem, they call an engineer. This is because an engineer understands the system well enough that they can create new procedures, thinking and working outside the box.

Other systems flight controllers are busy during nominal operations—for instance the Instrumentation and Communications Officer (INCO) is always busy tracking communication plans, switching and operating recorders, and sending commands for other flight controllers. They operate the systems directly so that the crew doesn't have to—that way the crew can concentrate on payload or other mission operations.

Trajectory flight controllers are always looking ahead to make sure that they know where the vehicle is going, based on where it is. They are tracking the vector, building burn and rendezvous plans, and—for ascent and entry—monitoring to make sure that the vehicle is going to get to orbit, or a runway.

Operational flight controllers are busy tracking the crew's execution of their plan as well as the execution of payload operators, who might be working remotely. The team of the Flight Activities Officer (FAO) is constantly keeping track of what has been accomplished and how much time it took, so that they can know what is left to be done and what needs to be rescheduled. Payload Officers are constantly in touch with the customer community to make sure they are getting what they need from the crew, the vehicle, and the ground team. The MCC is a busy place, and even when it seems quiet there is a strong undercurrent of activity. We always said that if you didn't have something to keep you busy, you were obviously missing something!

It generally took about two years for someone to become certified in a backroom flight control position. A new hire would start by studying all the basic material on the Shuttle as well as on their area of responsibility and how to operate in the Control Center environment. This might sound simple, but it was like drinking from a fire hose. All new hires looked at the mountain they had to climb to certification and were stunned. But you climb a mountain one step at a time. Before you knew it, a new hire would be sitting in on simulations and getting the feel for how things worked. In a few more months, they would be working the console position during simulations—as well as taking classes in the simulators for their particular area of responsibility and working on procedures, rules, drawings, and other documentation in the office. This office work not only reinforced what they were learning in sims but created new reference material for use on console. Another key element of learning was going to meetings—sure this sounds boring, but meetings were where flight techniques, rules, and procedures were ironed out—as well as where vehicle problems and changes were hashed out and argued over by program management, engineering, and the flight ops team. Meetings were where young flight controllers learned to understand logical argument supported by data—always supported by data! It was where they learned to sell their ideas to colleagues who were looking for any weakness in risk analysis and logical thought.

When a flight controller had reached a level of competence where their superiors felt they were ready for certification, they were examined by the section members who used a variety of tests—written and verbal—to check their knowledge. They then had to go through a certification sim where the script was specially designed to put them under pressure, and they had to pass not only technically but with poise and confidence. Only then could they be assigned to a mission—where the training would continue.

Front room flight controllers were trained in the same way, but to an even higher standard of abilities that enabled them to communicate and play in the give-and-take environment of the integrated space operations world. It might be painful for an Electrical Generation and Illumination (EGIL) controller to have to "ride a short" (watch a bus that is likely overloaded and smoking its wires), but if the bus was essential for keeping a critical piece of equipment operating to landing, then they had to bite the bullet and sacrifice their system for the good of the vehicle. It was common in the MMACS (Mechanical Maintenance Arm and Crew Systems) world to burn up an auxiliary power unit or to pump the hydraulic fluid out of a leaking system in order to maintain control. And nothing was as nerve-racking to an EVA (Extravehicular Activity) Officer as having to feed a suit leak so that a crewmember could button up a critical piece of equipment rather than leaving it exposed and scampering for the airlock. But if leaving the equipment exposed meant you had to do another EVA, well...there was always a risk trade involved. Oh—and if you think that these were the only operators (flight controllers were, for some reason lost to the mists of time, referred to as operators) who were nervous in these cases, remember that the ultimate responsibility always lay on the Flight Director's shoulders—and they were the ones asking the controllers to take these risks.

When it came right down to it, all the training, all the learning, all the education was merely building a library in each flight controller's mind and laying a foundation in their experience base so that they could make good risk trades. Controllers were there to execute, but they were mostly taught how to troubleshoot and manage risk. No decision was ever made in a vacuum (please pardon the play on words)—they were always made in the context of an immensely complex vehicle and operation, where even the smallest decision could have a rippling effect on everything that followed after. A minor change to a maneuvering jet configuration could impact flying procedures on rendezvous day, just as the decision to use up precious gas or cryo could affect the ability to lengthen the mission down the road.

We always trained people to think and work quickly, but at the same time to never make the right decision "too early." There is an old pilot expression that goes, "When something bad happens, the first thing you should do is to wind your watch." It means that instead of reacting instinctively to a situation you think you understand, unless the result is immediate loss of control or death, you should take a few seconds to think about what is happening and make sure you have properly identified what is going on. More often than not, you have more time to solve a problem than you initially think. Like pilots, flight controllers needed to be taught that being fast is not always the same as being right.

On the other hand, there are times when fast is the only way to react—and the most important thing was learning how to tell the difference between having time to think things through and needing to react immediately. One thing was for certain, no one ever got it right all the time, and no one could ever stop learning.

Chapter 4

Getting Selected— The Chosen Few

I don't know for sure when I decided that I wanted to be a Flight Director. I certainly knew of Chris Kraft and Gene Kranz from a young age, but I simply knew them as guys who had something to do with being in charge of space missions. I never really thought about a hierarchy in Mission Control, never really thought about it from an organizational sense. I guess it was obvious that someone had to be in charge. I think I was always enamored with the characters in the 1950s and 1960s low-budget sci-fi movies who were clearly the lead scientists in charge of saving Earth when aliens or natural forces threatened the planet and/or the human race. There was always some virile senior engineer or scientist who cut through the BS being spouted by world leaders and politicians to lay bare the problem and what needed to be done about it. People looked up to this leader who knew the answers and they followed him because of his knowledge, understanding, and charisma. To me, that was the essence of the Flight Director—a take-charge individual who knew the answers and inspired people to follow him.

But I am sure that when I first walked into the Control Center as a co-op student, I didn't recognize the Flight Director for who he was—that leader from the early sci-fi movies. I was introduced to backroom flight controllers and understood that they worked for their front room person—but at first, admittance to the front room was a rare and holy event.

Flight Director selections are an occasional thing. In the early days of NASA, there was one Flight Director—Christopher C. Kraft, the man responsible for setting the standards and establishing the culture of the job. Kraft came out of the NACA (National Advisory Committee for Aeronautics, which was superseded by NASA) flight testing tradition. He organized the original Mercury Mission Control around the concepts learned in atmospheric flight tests. Missions were put together around a set of objectives and a flight plan was built. The pilot and team trained around the mission plan, flew according to the plan, and debriefed what was learned—according to the plan. The flight plan was everything, and how well the mission held to the plan was often the criteria for success. The early Mercury missions were short. They consisted of suborbital lobs that lasted less than an hour, or short orbital missions that lasted a few hours and were conducted by a single flight control team—and a single Flight Director. As the later Mercury missions grew in length, additional Flight Control teams—and additional Flight Directors—were required, necessitating the first Flight Director Selection. The details are lost in bureaucratic history, but legend has it that Chris simply used his authority to ask his assistants—Gene Kranz and John Hodge—to train and take the other teams. The three of them decided that for the sake of scheduling and comradery, they would each pick a team color that would identify them and their crew on schedules. Naturally, the three of them picked Red, White, and Blue as those first colors, setting a team-naming precedent that is still used today (although teams are hardly fixed, and the idea is now simply used for the sake of tradition).

When the Gemini program came along, additional Flight Directors were necessary, with Chris stepping off console entirely to manage the much larger aspects of the growing program. In those days, selection was again a matter of appointment—good people were chosen to move up and take over more responsibility. As often as not, the "promotion" was done informally. There was no Flight Director Office until the beginning of the Space Shuttle program, and those who served as Flight Directors had regular jobs as members of Flight Control Division management.

Some were assistant Division Chiefs or Branch Chiefs, and their console duties were in addition to what they did in the office. It was quite common for section heads to serve on console in those days as well, the idea being that everyone had an office job but also served the mission in the Control Center. It was a concept that kept managers intimately involved with the day-to-day operations for which they were ultimately responsible.

The Flight Director Selection process continued to be informal into the early days of the Shuttle program. The first few Shuttle Flight Directors related that Gene Kranz simply dropped in to their office and asked if they were ready to take over as "Flight." Some may not have been given a choice. The Flight Director Office was formed up as an organizational entity about that time, and the active Flight Directors transferred into the office as a full-time job. The first true competitive selection that I recall was in the early 1980s. I distinctly remember several senior guys smiling and moving rapidly through the halls, filled with excitement at the fact that they had been given the nod. This was an early 1980s class—the men with whom I would serve as a front room flight controller in the years before and just after the *Challenger* was lost. These were the years in which I built some sort of reputation for being a good, common-sense systems guy who understood flight operations and the need to sometimes be quick rather than precise.

I know that I recognized that the way to be a Flight Director was to learn everything I could about everyone else's systems and operations—not just to concentrate on my own assignment and my own job. I took that idea to heart and studied everything I could. When I wasn't assigned to a simulation, and had nothing else preassigned to do, I would go to a back room and watch the sim—but not as if I were working in my discipline. I intentionally didn't punch up any of my discipline's loops or displays. Instead, I brought up the Flight Director Loop, the air-to-ground channels, and whatever display was on the Flight Director Console (we could tell where things were called up), and listened to the entire sim as if I were training to be Flight. It was an eye-opening experience, and I learned quickly how much I had to learn.

I first applied for the Flight Director position in 1989 or 1990. I was still relatively low on the civil service pay grade schedule, not yet a section head, but a recognized Ascent/Entry and Orbit front room guy who had fought a few battles of note. I had been the face of the Spacelab Instrument Pointing System at many meetings and boards before the *Challenger* went down. After that accident, I became one of the founders of a new MMACS (Mechanical Maintenance Arm and Crew Systems) position, and served as the Ascent/Entry MMACS for the Return to Flight with STS-26. My real claim to fame was probably in the landing/deceleration world, which I loved because it involved lots of flying stuff—especially in the big Vertical Motion Simulator (VMS) at NASA-Ames. I spent a lot of time with the Shuttle commanders and pilots for the first time (my Spacelab days were spent mostly with Mission Specialists). Again, I became known for a common-sense approach and an encyclopedic knowledge of flying machines.

I was having a great time as a MMACS, and the job put me in a strategic position to learn all I could about the whole Shuttle system and prepare myself for the Flight Director job. Since I started out in a bit of a backwater (the Spacelab Systems group), I had to work hard to get to know my counterparts in all the Shuttle systems disciplines. I did this by working on as many broad assignments as I could. The landing/deceleration improvements exposed me to a wide variety of rules and procedure discussions, and I spent a lot of time preparing for presentations and defenses of the new operational philosophies at many flight techniques meetings. It was a lot like defending an academic thesis for a graduate degree—flight technique reviews were freewheeling, with lots of very smart people trying to find holes in your arguments. It was a great place to learn the craft while making a reputation (for good or bad).

I looked for interesting side projects as well. For instance, one of the things that the program wanted the Mission Operations Directorate (MOD) to demonstrate before we returned the Shuttle to flight was our Emergency Mission Control Center (EMCC) capability. EMCC was an alternate to MCC (Mission Control Center) Houston—a facility that was vulnerable to damage (or worse) from hurricanes. The EMCC concept in those days was developed by the advent of a special portable computer that could decode the Shuttle telemetry stream and provide that decode data to laptop computers running special software that showed flight controllers what they needed to help the crew prepare for, and fly, an entry. All that we needed was a safe facility with access to the telemetry stream coming in from the satellites, as well as good communications capability. And someone figured out that the best place to do that was right at the ground site for the satellite antennas in White Sands, New Mexico.

The concept was simple—in the case of an imminent loss of the MCC, we'd load a team of flight controllers onto our KC-135 aircraft (the infamous Vomit Comet) along with their computers and a pile of checklists and flight plans. We'd fly them to Holloman Air Force Base on the edge of the White Sands Missile Range. There, we'd catch rides on Army helicopters across the small mountain range that separated the missile range from the NASA Ground Terminal, commandeer some desks and communications boxes from the ground terminal, and set up shop in our new digs. On paper, this looked great—the problem was that no one had ever done it. It was time for a road trip!

Flight Director Al Pennington was the leader of the "Great EMCC Migration of 1988," scheduled for October. We had a full team of flight controllers. Most of them were Ascent/Entry qualified and were assigned to STS-26, the first mission after *Challenger*, which we were getting ready to fly. We planned the activity to start with a simulation running in MCC, with a crew in the Shuttle simulator. We'd declare an emergency, and everyone and their gear would head for the airfield about 7 miles away, load up all our gear, and launch for New Mexico. The astronauts in the simulator would keep "flying" while we were in transit, and when we got on station, we'd make contact and control them from EMCC as they flew an entry.

This all worked fine, right up to the point when we were landing at Holloman. It turns out that in fleeing a make-believe hurricane in Houston, we'd flown right into a *real* blizzard in the high desert! After a temporary diversion to Albuquerque, while Holloman plowed their runway, we finally touched down (with some slipping and sliding on the remaining snow and ice) a couple hours late. We then loaded up the helicopters and waited for good enough visibility to launch—which finally occurred after noon. The flight across the highly classified missile range was interesting as we were encouraged "not to look out the windows, and to forget anything we might see." Over the mountains, we approached the ground terminal antennas bristling with fresh snow shining from the ground. Our landing site was the approach road to the facility, and with six Army Hueys making combat-like approaches, we looked for all the world like a scene out of *Apocalypse Now*. Trucks and vans met us and our gear, and we were up and running within about thirty minutes of arrival.

It was only after the fact when we realized that about 80 percent of the Ascent/Entry team for STS-26 was participating in this adventure, and when one high-level manager asked, "What would have happened if you'd had a helicopter mid-air in all that snow?" we realized the programmatic risk we'd taken. If we'd lost a bunch of key people in an accident, we could have delayed the Shuttle flight by a year or more while replacements were trained. It was decided that from then on, if we ever did an exercise like that again, we'd make sure we took less critical players... just in case. But it sure was fun, in addition to being educational!

We trained harder and longer for STS-26 than for just about any mission NASA ever flew. We had the time and we wanted to stay current, so the hours piled up. We flew so many cases that they all blurred together, and we finally felt that there was little more the sim team could do for us—or to us. It was a great team to be on, and most of us stuck together in later years as we went on to lead MOD in many different positions.

Despite the fact that I was a grade lower than most applicants (most folks applying were GS-13s with a section head job under their belt and were looking for a promotion to GS-14 that came with a Flight Director selection), I was encouraged to apply when that 1989 selection call came along. Apparently, I had impressed a few people along the way as a flight controller, and there was always an informal (probably unwritten) list of flight controllers who were being watched as possible candidates. I spent hours working on my application, trying to get the words exactly right so that I would be among the relatively few who would get an interview. Selections took a long time back then—a half a year from the first announcement to the final selection was not uncommon. And there were all sorts of tallies kept on whiteboards throughout the flight control offices: who had applied, who had gotten an interview, who had been asked back for second interviews, and so on. It was a topic of conversation, and those in the game were usually the ones not talking.

When I got a call from the secretary in the Flight Director Office to schedule me for an interview with Mr. Larry Bourgeois, that was a pretty big deal—just to get an interview, and to get it when I was so junior. I had not yet been a section head, and the section head job was considered a prerequisite simply because it ensured that candidates had received a certain amount of personnel management training and experience—rudimentary leadership credentials, if you will.

If I recall correctly, and this may be colored in my memory, I think that there was a somewhat illicit list of interview questions from past selections floating about that I used to prepare for the interview. I know that for my second and third interview tries (in later years), I had my first interview prep to go by, but I remember preparing fairly thoroughly for the first, so I must have had some idea what would be asked. That first interview was one-on-one, just me and Larry. I think that we had worked together on console once or twice, but not a great deal, so we didn't have a lot of background—yet the interview was thorough and collegial. I honestly think both of us knew that it was a rehearsal for a later time. There were good, solid candidates who were in line for this selection, and it would have been astounding for me to get it. I saw it as a learning experience, and it was—a good way for me to get an idea of what they were specifically looking for, and a way to calm my nerves by exposing me to the process.

Interviews went on for several weeks, people's schedules being what they are. When the interviews were complete, the waiting began—and it went on, and on, and on. Back then, the Office Chief made his selections, passed them up the line to the head of Mission Operations (Gene Kranz) who, if he agreed, passed them up the line until they reached some fairly lofty position at Headquarters. Flight Directors are the men and women in the hot seat during human spaceflight missions, and the administrator needs to know (and have faith in) whomever is holding the stick.

Tradition has it that those candidates who are *not* selected get the word first—usually with a phone call from the Flight Director Office secretary, who informs them that the chief wants to meet to debrief their interview. This call meant that you didn't get it—sorry, but we'll tell you what we thought about you. As phone calls went out, and debriefs were scheduled, names were crossed off lists on the whiteboards throughout the building. Finally, only a couple names were left—and the selectees were known. I, of course, was not one of them. I felt sort of like the loser in a raffle drawing. I had no real expectation of winning because the odds against me were too great, but sure, there was a little disappointment that I was going to have to wait a couple years before the next chance. In the meantime, of course, I had a lot to learn—and the MMACS section head job was calling.

And it appeared that I was in line to get that section. Our long-time head was moving on, and I was the senior MMACS Officer, so it seemed like a shoo-in. Unfortunately, it wasn't quite that simple. It turned out that we didn't have the depth in the section that we needed, and if they gave me the job, I'd either be a part-time section head because I was needed on console, or we'd be shorthanded on console. The bottom line was that the job came open just a little too early for me, and so I didn't get it. But I was told to be patient. And sure enough, less than a year later, we had improved our staffing situation and they moved my predecessor out (to another section), and me in. During that time there was another Flight Director selection, and while I once again went through the interview routine, I was once again told, "Not quite yet—get a little more time in the section head role." When I asked what I should do differently for the next selection, I was basically told, "Don't change a thing." It just wasn't my turn.

An interesting thing happened between that selection and the next—I became a noted expert in Russian space systems. At least, I was noted that way within our organization. I had always been curious about the Russian space program, but not in a way that I went after information. To me, it seemed that what they did was fairly primitive, as were their systems. I was much more interested in the things we had going on in the Shuttle world, and particularly in trying to run the largest section in our division with a certain amount of efficiency. The MMACS section was comprised of two major groups—the traditional systems flight controllers for the Orbiter's mechanical systems comprised one half, and the experts in Crew System and Inflight Maintenance (IFM) made up the other side. This second group did much more than flight control. In fact, flight control was a small part of their job. They primarily oversaw crew training and operations engineering for all of the many things that occupied the crew cabin—cameras, clothing, seats, suits, the galley—as well as all the tools and maintenance procedures with which the crew needed to become proficient.

The size of the section was such that it was really too big for one guy to handle—twenty-nine people at one point, counting both contractors and civil servants. It was roughly twice the optimum size and, of course, there was no such thing as a deputy section head. In order to manage things, I picked my most senior mechanical guy and my most mature IFM guy, and I asked each of them to lead their respective halves of the section. I told them there was no more pay, more responsibility, and that they should consider it a good training ground for what might come later on. In other words, I delegated. To me, showing trust in people is the best way to get them to perform beyond what they think they can do. And in this case, it worked well—so well, in fact, that I found I had extra time on my hands.

I was known throughout our division management as someone who preferred technical assignments to managerial assignments; if I had a choice, I'd take on a technical evaluation long before I volunteered to come up with a new management process. As a result I was asked to do interesting things. When the new Shuttle *Endeavour* was being readied for shipment from its assembly plant in Palmdale, California, to the Kennedy Space Center, a team of senior NASA people from Johnson Space Center (JSC) had to go out and perform a Management Acceptance inspection. Jack Knight, then head of the Systems Division, asked me to go because he knew I worked on airplanes and would have a decent chance of finding any issues with assembly or finish work. This turned out to be a great trip. We crawled all through the Orbiter—I even took a caving excursion deep out into the left wing tip. A hatch inside the left wheel well led into the internal bays of the wing. The ribs weren't built the way you would see in a typical aircraft, with solid metal stampings cut through with lightening holes. Rather, the ribs were made up of individual composite struts acting as a truss to keep the upper and lower skins apart. I had to work my way through the maze of struts, looking at the quality of the work while being extremely careful not to put the slightest side load on any of them. These were paper-thin boron tubes that depended on their perfect cylindrical shape to maintain compressive strength. It was a challenge—much like moving through a cave passage with delicate speleothems (formations) on all sides, none of which you are allowed to touch. Yes, I was glad I had several years of caving experience back in college—and I was probably the only person in the entire organization who did.

The Palmdale trip was interesting, and after it, in early 1992, I got a call from Flight Director Gary Coen asking if I would be interested in taking a special overseas trip. Gary had been the Ascent/Entry Flight Director for STS-26, the Return to Flight mission after *Challenger*, and I had been his MMACS Officer. That team trained more hours for that flight than any other Ascent/Entry team ever had, and we grew very knowledgeable about each other. It seemed that Gary had been asked to put together a small team of operations experts to answer a Headquarters call for a trip to Moscow. The Soviet Union had broken up just a few months earlier, and shortly thereafter President Clinton and President Yeltsin had gotten together to talk about many things. One of those things was high technology space workers, and what was going to happen to them in a country that quite frankly had no way to support a space program anymore.

The concern of our government was that these space and missile experts might end up selling their services to the highest bidder for the simple reason that they needed to survive and feed their families. Those highest bidders were countries that wanted to develop missile and weapons capabilities—nuclear weapon capabilities. Our counterparts in the

Soviet (now Russian) space program were experts in ballistic missiles. Having them go to work for countries unfriendly to the West was clearly not in the best interest of any nation in the Western sphere of influence. The solution was simple—we needed to give the Russian aerospace community something to do, and the money to do it with. And so was born the idea of cooperation in space. The presidents turned it over to their respective space program heads to figure out the details, but the basic idea was that we needed to come up with something that we could do together in space that would allow the US to contribute money in their direction. The Russian program cost a great deal less to operate than the US program (due to the low valuation of their currency), so a small contribution from our side could make a great deal happen on theirs.

Of course, the heads of the two programs met and talked about cooperation, but they quickly turned it over to their experts to actually figure out what we could do—what kind of a joint program we could put together. The Apollo-Soyuz test program of the mid-1970s had docked two spacecraft—one from each country—in low-Earth orbit as a symbolic gesture. So naturally, we thought of docking. Gary called me because he was taking a small team over to Moscow to discuss the potential for joint missions, and such a mission would need a docking system of some kind. I had been doing some technical evaluations of docking and berthing systems that were being developed for the Space Station Freedom project (another of my "interesting technical assignments") and, therefore, I was the most qualified person he could think of to go along as his systems expert. It helped that I still had a valid official passport from my Spacelab days, which meant that I could be ready to go quickly.

That first trip to Russia was amazing. We all had the same feeling— we were headed into the enemy camp. Everyone on the small team had grown up in the 1960s with the imminent threat of nuclear confrontation hanging over our heads. We talked about that, and without an exception, we had all grown up expecting that the world would end in a nuclear confrontation. Russia was the enemy—an unknown and dark enemy who couldn't be trusted. Russia was a mystery, made more mysterious by the fact that we didn't speak their language, and they often didn't speak ours. Interpreters were the rule. We'd all been raised on East-versus-West spy movies and we suspected that there would be KGB agents watching us every moment—would they suspect us of espionage and whisk us away to some basement prison cell? It was an uncomfortable feeling, but it proved to be an adventure, and our fears were unfounded. The Russian space people were just like us in the end—and not, as those of us raised on *Star Trek* assumed, like representatives of the Klingon Empire.

On our first few trips to Russia, we always stayed in a fancy, modern, Western-style hotel built specifically for European and American guests—the Pentahotel. Since I was used to staying in Holiday Inns or Best Westerns on official travel, this place was amazing. I'm sure anyone who travels extensively to large cities would have just found it to be normal accommodations. I believe that we were told that Lufthansa owned it at the time, and all the food and drink were brought in from Germany every day so that the cooks could work with the finest-quality foodstuffs—not what was available locally. Aside from the lavish fit and finish of the place, I was impressed that there were large, fit men in black suits with little ear phones and coiled cords going down into their collars stationed in the lobby. I mentioned this to my translator one day, that I thought it was interesting that the government was supplying security for us. "Oh, *nyet*," he said, "those men aren't from the government; they are from the Russian mafia. They are here to make sure that the people with money, foreigners, aren't inconvenienced by petty criminals or locals who might harass them." I thought it odd that our security was being ensured by the dark side of Russian culture, but I was pretty sure that no petty thug was going to take a chance messing with us.

The building where they drove us for our first meetings was something that we would refer to as an "off-site conference center." We drove for quite a while getting there and, at one point, we passed a sign that clearly showed a pictorial representation of a shooting range in one direction, and an office building in another. We were just happy that they turned toward the office building the first time we went past it. We still sometimes referred to

the place as the Rifle Range, though. We had big meetings in lavish conference rooms, and if we asked for someone to make a copy of a piece of paper
so that everyone could see it, that was a big deal.

Apparently, they had only one copying machine in the entire facility. We later found out that they had one, almost original, IBM PC in a
little office—and that was all the computer power they had to support
our meetings. We soon learned to bring laptops—but always special ones
we checked out just for Russian travel. There was concern that we might
bring back viruses (which did, in fact, happen), and they wanted to be
able to wipe the laptops clean and keep them isolated from the rest of the
JSC computer network. Later on, when thumb drives became popular,
because of the threat of computer viruses we were warned to *never* put
a thumb drive we obtained locally into any computer we had. This was
a wise piece of advice, and a best practice that was proven to be correct
more than once in the years people went back and forth.

By the time spring of 1993 rolled around, I had been to Moscow two
or three times. Space Station Freedom was stalled and the funding future
was cloudy. Freedom had survived a funding battle in Congress by something like one vote, and the writing was on the wall that it wasn't going
to survive forever if its only purpose was science and engineering experi-
mentation in low-Earth orbit. And so, as always happens, politicians got
involved. While we had been diligently working to prove that, yea verily,
we really could dock an American Space Shuttle to a Russian space station, forces in Washington were busy coming up with a strategy that was
bigger than that. Freedom was to be rescoped and redesigned—that was
a given. Virtually the entire program office in Reston, Virginia, was reas-
signed or let go—the program was effectively over. Freedom was done—
but the idea of a space station was not. A new set of offices was set up in a
tall building in Crystal City and a tiger team was created. Their purpose:
come up with a space station that we could afford, an International Space
Station that included the Russians.

As luck would have it, the spring of 1993 saw an announcement that a
new Flight Director selection was going to take place. I once again pulled

out my folder of application paragraphs and interview questions and submitted my paperwork by the deadline. Based on previous selections, we figured that interviews would not be until late summer or early fall. In the meantime, we had to keep busy doing what we were doing, while waiting for the wheels of the process to turn slowly. Because my unofficial deputies were doing such a great job taking care of the section on day-to-day business, I was able to spend my extra time looking at the potential docking mission with *Mir*, the Russian space station. And then I got a phone call from the Directorate office, asking if I could head up to Crystal City for a few weeks. Crystal City? Hmm...I had been studiously avoiding work on the space station for several years—why did they want me to go to Crystal City, home of the "Freedom Redesign"?

Well, I was told, they were going to be bringing in the Russians for some preliminary meetings, and I was the closest thing we had to a Russian system-operations expert at the time. Besides, they had asked for a Flight Director, and they had thought of me.

Hmm...what the heck did *that* mean? I wasn't a Flight Director. I had simply sent in my application! I mentioned that fact on the phone, and I was told again they needed someone who could represent the Flight Director Office, everyone in the office was busy, and they wanted me to go. It would be four months before the selection process was over, and having been disappointed by unhatched eggs before, I didn't really want to think that I was a shoo-in, but I took it as a positive sign and headed for Washington.

When I got to Crystal City, I found a typical tiger team situation—lots of people living in hotels close to the office space, sixteen-hour days, schedules that assumed everyone was available at all times of the day and night—and a very troubling sign on the door of the office they had assigned me. The office itself was pretty nice, in that it had a view that stretched from the Pentagon out toward the National Mall, and we were way up high. It was nice. The troubling sign? My name was already engraved on a plaque—not just written in marker on a piece of paper. This was a bad sign in my book. I had planned to come to Washington for a week or two of meetings with the Russians—I had not planned on staying! Working in Washington has never been a goal of mine; in fact, quite the opposite is true. Oh, it's a nice place to visit—but live and work there? I'm sorry, there really isn't any kind of work that I enjoy that is actually *done* there. It didn't take long until I was on the phone to Randy Stone, the head of MOD, telling him that while I was happy to stay for the Russian visit, I wanted to make sure that he was ready to get me the heck out of there! Whether that would hurt or help my chances of being part of the office, I really didn't care. All I knew is that Washington was not for me.

The meetings with the Russians were interesting because I knew them and they knew me. Not well, mind you, but I was a familiar face, and they understood who I was and what I knew. That made me a person who they listened to, and in small sessions it seemed I had more clout than I officially did. The Russians came to Crystal City with a sound idea of how they could help the effort to get the new International Space Station kick-started and off the ground—and what they proposed turned out to be very much where we ended up. Give it to the Russians: while their equipment might seem primitive, they generally have simple ideas that work. And after we have run around trying to come up with complex solutions, cost and scheduling realities often drive the program right to where the Russians started.

The summer rolled on, and I was rescued from Washington. I continued to spend much of my time working Russian issues with the small group within MOD that was looking at the docking mission. There was a lot that needed to be done to make it work—not the least of which was modifying an Orbiter to handle a docking system. That was Engineering's problem. Ours was figuring out how best to rendezvous and dock without blowing the solar arrays off the *Mir* in the process. We held regular weekly teleconferences with the Russians, each one starting early in the morning our time and late in the afternoon for the Russians. I guess that from 1992 on, I lived on a global clock, working hours and shifts as necessary, but rarely working a week of normal hours.

By late October the Flight Director selection process was still not wrapped up. Most of us involved had reached the point where we were fed up enough (almost) to the point of not caring—so long as it came to an end. In early November, we held yet another set of joint meetings with the Russians, this time on our turf. Because of the nature of security and bureaucracy, we held these meetings off-site, just out the side gate of JSC at the leased Regents Park building, a nondescript two-story, smoked-glass structure where we could reconfigure rooms to meet the number of splinter meetings we had scheduled. There was always a meeting or two going on, but most of us were only needed about half the time, so we'd drift in and out of empty offices and back and forth to our offices on-site when we had an hour or more. It was a different, separate world from our normal jobs, and time seemed sort of suspended. Every day seemed the same—the topics were the same, the arguments the same, the details of going over protocols . . . the same. Translation followed translation, and it seemed like every day we started and ended at the same place; progress was slow enough so as to be hard to measure.

It was about this time that the operations people working the project adopted the movie *Groundhog Day* as their theme. The movie featured Bill Murray as a news reporter stuck in a small town, forced to live the same day over and over again—until he learned the key to life and happiness and broke out of the time warp. We had no such lesson to learn, and things just kept on going slower and slower. Coupled with this was the knowledge that the Flight Director selection process was actually supposed to be coming to a close, which made things even more excruciating. Rumors about names being crossed off whiteboards were flying around, but it was all happening a half mile away, back in our on-site office, while I was trapped at Regents Park.

What we were waiting for was "the call." We didn't want to get it too soon, of course. But without a copy of the scorecard, I didn't know where they were in the process. Thanksgiving was coming. In fact, it was the day before Thanksgiving, and we were trying to wrap things up with the Russians so that everyone could have a long weekend. I couldn't believe that we were going to have to suffer through four solid days of limbo—knowing that we hadn't been dismissed, but not knowing who had been chosen. I think the holdup was that some of the nonselectees couldn't be found. Well, of course not—it was a holiday week! I distinctly remember leaving the Regents Park conference room the day before Thanksgiving with no call, and telling myself that I just didn't care anymore. I really had given up. And then—one of the secretaries assigned to the building handed me a note saying that Lee Briscoe wanted to talk to me. Lee was the Chief of the Flight Director Office. And he actually didn't want me to call; he wanted to know if I could be in his office at 2:00 p.m. that day.

Nonselectees didn't get a call asking to be in the Chief's office at a particular time; they got a call from the secretary to schedule an appointment for their debrief. It was noon when I got the note over at Regents Park, and I was fairly certain I knew what it meant. I also knew that folks back in the office had been keeping score on those ubiquitous whiteboards, and I really didn't want to run into anyone to talk about what everything meant until I knew for sure one way or another. Seeing as how it was lunchtime, I figured that the best chance not to run into anyone was to head back to the office right away, then hide out with my door closed. To even further avoid folks, I didn't walk the long way through the building to my office (which was on the far end from the parking lot)—I walked around the building and came in the short way. I later found out that a couple of my friends had seen me, and, noting how absolutely intent I was on staring straight ahead and not pausing, they had a pretty good idea what was up.

I got to my office an hour before the meeting, and I can't for the life of me remember what I did for that hour. The WWW part of the internet hadn't yet been invented, so I couldn't have been surfing the web. I probably just stared out the window and waited for the clock to come around to the appointed hour. When the time came, I headed up to the third floor, corner office, and when I came out of the stairwell, there were Flight Directors standing in their doorways, smiling and looking at who was headed to see Briscoe. It was then that I knew for sure—I was getting my chance to prove I had what it takes to sit in the center seat.

Chapter 5

Becoming Flight

I walked down the short hall to the office of the Chief Flight Director, and when I turned in there, I saw two other faces that I recognized— John Shannon, of the Guidance, Navigation, and Control (GNC) section, and Bryan Austin, a Simulation Supervisor. I knew John—we'd flown on teams together, but Bryan I really hadn't spent much (if any) time with. As we entered the secretary's office, there were handshakes and congratulations. It was clear by then that we were the three lucky selectees to make it through the process. We were the Class of 1993.

Sitting down with office chief Lee Briscoe and his deputy, Gary Coen, our joy was short lived. Lee began with something like, "Gentlemen, congratulations—you may have been the cream of the crop, the top picks within the Directorate, but now that we have you here in the office, you're lower than whale shit on the bottom of the ocean." Okay, so it didn't really dampen our euphoria that much; we still knew that we were now on the road to the center seat. But we also knew that we had a lot of work ahead of us, and part of that work was to prove to the office and to everyone who worked in MOD (Mission Operations Directorate) that we were worthy of the job.

Logistics came first, of course. Transitioning to the office was not an overnight job. Frankly, the fact that the selection ended up at Thanksgiving gave us about a month between that holiday and the Christmas holidays to get ourselves extricated from our previous roles and move up to the third floor so that we could effectively start fresh in the new year.

December has always been a lonely time in the MOD buildings. Unless we had a reason to be flying a mission, it was rare for anything important to be on the schedule. This was mostly because we all worked so hard most of the year that everyone needed a break. It was also a good time for folks to use up their accumulated "use or lose" leave and comp time. With lots of people gone, it was a quiet time for us to pack our old offices and move to the new ones—three freshly equipped rooms next to each other on one end of the Flight Director Office. I can't remember who picked theirs first, but it might have been me—the only differences between them was who had the thermostat that controlled all three rooms.

Previous Flight Director Selectees had shared an office during their year of training, usually because the office didn't have that much room to spread out. We, however, were blessed (or cursed) with the fact that the office was expanding and had moved into the space previously occupied by the Astronaut Office (which had been located there since the founding of Johnson Space Center). The blessing was that we all had our own private office—for me, one I would inhabit for the next twenty years. The curse was that we were all going to be training together, and a common room could be advantageous in terms of training and reading as a team. I don't know for sure which model is better (subsequent classes shared offices at times, and some liked it, some didn't). I personally liked the privacy for reading and researching—plus we could always all go to one office for briefings and study sessions.

That first six months was designed as academic study—we needed to get up to speed on all the systems that we weren't expert on, and we had to do it quickly. We did this by reading the training workbooks, the operations manuals that gave the details of the procedures, the procedures themselves, the malfunction procedures, and the rules. We had briefings by the systems experts (many different ones every week). We finished up by doing the crew lessons in the part-task training devices and simulators. Finally, we'd have a briefing from each discipline on "rules of thumb" and other tricks they had that weren't really in any of the other documents. For my pilot friends, I described this period as the Type

Rating from Hell—we had to learn everything there was to know about the spacecraft so that when a flight controller told us about a problem, or asked us for a procedure, there was no mystery behind the request.

The tricky part about being the Flight Director is integrating the knowledge of all of the various disciplines, thinking about how everything fits together to accomplish tasks. No matter how much we emphasized integration, and no matter how well our flight controllers worked with other disciplines, there was always a bigger picture that only the Flight Director really held in their head. After all, we had the ultimate responsibility for mission success and crew safety. The big picture was ours. Someone once described the Flight Director as the conductor of the orchestra: we might not know how to play all the instruments well, but we could probably make noise with any of them—and we knew what they *should* sound like when played well.

It was a lot of work doing the academics, even with our long and extensive backgrounds of interfacing with all the disciplines. Nevertheless, we had a lot of fun too. For instance, we took several tours of various centers. Seeing as how the Flight Director needed to interface with all the NASA Centers (and Headquarters) at various times, we planned trips to go and meet the people and organizations, check out facilities, and let others know who we were. John, Bryan, and I took several trips to the East Coast sites, hitting Headquarters and Goddard on one, Kennedy and Marshall on another. For some reason, we never did make a good West Coast trip that I can remember. But I had spent a lot of time at Ames flying the Vertical Motion Simulator, had visited Dryden for landing and deceleration work, and of course, the Great EMCC Migration of 1988 checked off my square for visiting White Sands. I think I'd been to Downey as a fill-in representative for a program review one time. Nonetheless, we did our best to hit most of the highlights of Shuttle centers.

Headquarters was an interesting tour. We had a visit with the NASA Legislative Liaison—essentially NASA's chief lobbyist to Congress (although such a position title would have been illegal). We were interested in the fact that the year before our selection, the Space Station

Freedom program had survived in the House by a single vote. The year
we visited, the new ISS had survived by what could only be called a
landslide. We asked the Liaison (whose name I don't remember) if he
could give us an idea how many of the votes that had been "swung" were
changed because the ISS, and space exploration in general, was the right
thing for the nation, and the right thing for mankind. "Oh, that's easy,"
he said, "the answer is none!" That was the purest view into the political
legislative process I have ever had. Not one vote was changed because
of a moral imperative for the future of mankind—the changes were all
for political reasons, many of which were not obvious because of the
amount of vote trading that goes on. A congressman doesn't really support something, but votes for it in exchange for someone else voting for
something he wants but can't be seen supporting. It convinced me once
and for all that Washington was not the place for me!

Another enjoyable aspect of our academic training was a little
something I cooked up with my old section mates. I felt that if we really
wanted to be able to understand and empathize with the crew, we needed
to take as much of the crew systems training as possible. In other words,
the training that astronauts go through. We needed to know what it was
like to suit up, un-suit, use the galley, work the camera gear—and go to
the bathroom in space. Crew systems training and in-flight maintenance
training were always fun. While I had done it all in the past, I figured
that doing it again would get us out of the office once in a while and
allow us to get our hands dirty. It's funny that it can be the mundane
things that a crew has to do on orbit—like cleaning filters or cleaning
the toilet—that can make or break the quality of life on board. And make
no mistake, these small things have a huge effect on morale and performance. Knowing what they are going through when we asked them to
do these tasks—and why a radio response might be delayed because they
had their hands full of floating garbage—would make us better able to
understand what we could expect from the guys hurtling around the
planet.

I always enjoyed suited crew escape training, and I was happy to go through it again. For John and Bryan, it was a first. I had helped out as a test subject (that is, I was a guinea pig) during the development of the crew escape systems in the post-*Challenger* time frame. Because I was going to be responsible for the systems as a MMACS (Mechanical Maintenance Arm and Crew Systems), I wanted to be able to understand the systems from the inside out. It also reminded me a lot of my days as a professional diver—encumbered by lots of heavy gear that restricted movement and made you hot. This training would provide good insight into understanding what was possible and what was not. I was already fitted for a suit, but John and Bryan had to go through the measurement process out in the Boeing building to get ready for these exercises. NASA didn't build a suit for each astronaut—rather, they had several different sizes of suit that could be fine-tuned for leg and arm length, limb diameter, and torso fit using lacing cords. A suit fitting consisted of a session where they picked your initial size, and then tailored it to fit using the various adjustments. When they were all done, and you had good mobility, they'd write down everything they needed to know to make the suit fit you well. When your name came up for a training exercise, they'd put one together using your "spec sheet," seal it up in a box with your name on it, and deliver it to the training site. These fittings included choices of things like glasses holders, kneeboards—all the equipment that was personally chosen by an astronaut for their ascent and entry kit.

The Suited Ascent and Entry egress class was a great example of the expense that NASA would go to for training. John, Bryan, and I dedicated a day in the mock-up facility. This meant we needed suits made up to fit, and each of us had a suite of technicians who basically followed us around for the day. (Well, actually, since they were responsible for those expensive suits, I think they were following the suits around—we just happened to be the squishy parts inside.) In addition to the suit techs who got us in and out of the orange bags, there were the mock-ups themselves. In this case, both a Crew Compartment Trainer (CCT) and the Full Fuselage Trainer (FFT) were needed for the class. The CCT had to be put in its vertical Launch position. This took a number of hours of prep and handling time to make sure that when it was tilted to the vertical position, nothing went "clang" inside. Meanwhile, a Flight Data File had to be prepared and installed to mimic the launch configuration. As for the FFT, it had to be set up for Entry and Landing configuration, so that we could practice bailing out, and also escaping out, of the overhead window after a presumed intact (and upright) crash landing using our rappel devices. Oh yes—we also had to have a crane and gym mats dedicated to us in order to practice using the rappel devices before going out the overhead hatch.

Such a session was representative of the kind of training that we went through. Our goal was to *know* what the crew was going through so that we could *know* every aspect of procedures and the consequences of our decisions and actions. Getting out of the pilot's seat in full gear on the launch pad—or during a gliding bailout—was not a trivial task. Sure, you could get that idea by reading or watching, but nothing equaled the experience of actually having tried and done it. We got to strap in for launch, then we tried to get out of the Orbiter in a hurry after the Launch Control Center (LCC) or pad leader declared an emergency. We got to go down the emergency escape slide after "landing" at a runway that was not prepared to handle us. We got to hang from an escape rope and descender that was hooked to the building crane and hoisted 20 feet in the air. Fortunately, we had a gym mat to break our fall if we did anything wrong. We got to experience that line in our job descriptions that said, "Occasionally, physical exertion is required in training devices and simulations."

Wearing a Launch and Entry Suit is a great lesson in patience. Just like the experience I had with a diving dry suit in Minnesota, any extra effort was not useful. It just heated up the suit. If you weren't required to do a task, you quickly learned to simply stop moving. You consciously decided not to move your arms or legs. Even moving the head was a deliberate act. After suiting up, I'd move my head all the way around and try to touch my chin toward my shoulders in each direction to make sure that the wires to my comm cap (inside the helmet) weren't restricting my motion. But then—I'd become still and wait for each task.

In the early years of the orange Launch and Entry Suits, the crew wore heavy Patagonia underwear that was designed to keep them alive in the event of a North Atlantic bailout. There was not, however, much of a way to keep the crew cool—a shortcoming that was discovered while training in the heat of a Texas summer. It didn't take long for that to be rectified. This was done with the addition of water-cooling tubes that were sewn into the underwear, and the use of chilled water circulation. During our training exercises, we didn't have the cooling units that were normally attached to the crew's seats. As a result, we would still get warm during the exercises in the cockpit if we moved around too much. When we got outside for a break, the suit techs would be waiting with these big coolers filled with ice and circulating pumps that had been scavenged from an aquarium. These pumps circulated water through hoses that could connect to the suit's cooling fittings. After an exercise, while we waited for technicians to set up the next scenario, we'd sit on folding chairs next to the ice chests, hooked up and cooling down.

Getting out of an Orbiter on the launch pad was always a bit surreal. We spent so much time training in simulators with the Orbiter in its horizontal position (like an airplane sitting on the ground), that when it was tilted to point at the sky, you'd swear that it was a completely different space. You had to think hard after sliding out of your chair: "Which way is the hatch?" When you climbed through the "floor" (now a wall), which way did you turn to find the way out? This training gave us a good appreciation for what the crew might have to do on the pad in case anything went wrong—and made us certain that we didn't want anything to go wrong. Of course, this was all the responsibility of the launch team at Kennedy Space Center, but knowing what our colleagues in the orange suits had to go through was important. Although we would never actually feel the jolt of the Shuttle leaving the pad, we wanted to have as much of a feel for what the crew had to go through as we could.

Training was not without its hazards. In the years I was involved, I know of at least two people who broke ankles while taking the ride down the escape slide. You had to remember that all that equipment weighed quite a lot, and it wasn't like going down a playground slide as a kid in grade school. You hit the bottom like a ton of bricks. Plus your motion was significantly restricted, making reaction times slow. The Crew Systems world had a special set of helmets (one size fit all) set aside just for these kinds of ground training exercises, because they knew they'd get beat up during this training. Escaping out the top of the Orbiter and using the rappel device to lower yourself to the ground was an exercise full of bumps and bruises. There was no way to be graceful about it (like a rock climber). If the crew had to do it after two weeks in space, we always envisioned there would be a pile of inert bodies at the bottom of the ropes, laying there and waiting for rescuers to come pick them up.

Getting an appreciation for the day-to-day lives of a crew in space, despite the difficulties, was a lot of fun. But the hours spent in this type of training were simply piled on top of the pure academics, making for long days while we prepared for the second phase of our training—the time in MCC (Mission Control Center).

This second phase of our training started as soon as we had enough of the basics down from our academic studies to not embarrass ourselves. After six months out of the Control Center, we were about to make the MCC our home for three out of five days every week. And for a Flight Director, the MCC—and the center seat—is where the rubber meets the road.

The second six months of our training year included more study, but most of our time was spent in the MCC doing simulations. We took full advantage of the valuable sim time, always cognizant of the fact that hundreds of people were spending their time running the sims, supporting as flight controllers, and running the buildings. There were generally three generic orbit sims per week at the time—and since there were three of us in training, each of us got one. But that didn't mean we only had one day in the MCC—no, it meant that one day we were in the hot seat, and the other two days we sat behind so that we could watch our compadres and evaluate their performance. You could learn almost as much by watching as by doing. In some ways you could learn more, because you could see the big picture a little better. It was a great time, to be honest—we cut our teeth on Spacelab's Instrument Pointing System (IPS) mission simulations (because that was the generic load available) and doing rendezvous. I'd bet that a third of the sims ended in emergency deorbit cases—something that pretty much went away in later years, as the scripts for generic sims became what we called "kinder and gentler." Folks complained that when the sim team piled on too many failures, they couldn't take time to learn the actual procedures—they mostly had to resort to broken field running. Ultimately, the powers that held sway dictated that trainers could only work us two failures deep.

The fact that the Shuttle was designed to pretty much absorb two failures and still be operational made it hard to get into really nasty cases—but not always. The really good Sim Sups still knew how to rattle our cages and put us in precarious spots, and that was actually where the fun lay. We missed those cases in later years—the job became less of a challenge, less of a battle of wits with the sim team, and more of a "fill the square" exercise in trying different canned procedures. But that was all in the future for us during that first summer of training—John, Bryan, and I took the full force of what the sim team could dish out, and we had a ball.

One of the funny things about becoming a Flight Director was that you could always tell where a new trainee had come from. As a rule, when the going got tough folks tended to run home to what was familiar to them. That meant they'd spend more time working problems in their old disciplines, the things they understood better. The fact of the matter was, new Flight Directors were a pain in the butt to their old disciplines, because they almost always knew more than the poor trainee sitting on the console they came from. Those trainees couldn't catch a break. The Flight Director trainee might not know anything about how other disciplines systems worked, or how they handled problems, but they could sure show their old discipline a thing or two! So it was easy to see where the selectee came from—except in my case.

I was, at the time, a rare bird. This is because I had switched flight control disciplines after moving into the front room. Having started with the Spacelab Command and Data Management System (CDMS) and the IPS, I moved over to the MMACS group after the *Challenger* accident. So folks expected me to ride the MMACS position pretty hard. But in reality, you never know your second (or third) discipline as well as you know the systems you grew up with. I had learned how to be a MMACS operator from the guys who built the procedures and rules in the early days—I didn't build them myself. I had never spent any significant time in the MMACS back room, and I depended on good MECHs (the MMACS backroom position) to be an active part of the team and feed me what I needed to know. Although I was pretty good at being a MMACS, my real depth of knowledge was in the Spacelab CDMS and IPS world. And that, coincidentally, was what we spent our time with during that first summer of sims.

Now the funny thing was, most of the folks training as flight controllers at the time that we were training to be Flight Directors were young enough to only remember me as a MMACS operator—they really didn't know about my old IPS position. In fact, the IPS group had been absorbed by Guidance, Navigation, and Control (GNC) after I left, and the console position had changed considerably. The systems hadn't changed much, and neither had the procedures or rules, so they were very, very familiar to me. We did activations, pointing, and deactivations. We had problems with getting it up and operating, problems with runaway gimbals, and problems stowing that led us to jettisons (frequently). During all of this IPS work, we had the usual problems with Shuttle systems, and I was often complimented on the fact that I wasn't riding my "old discipline" (the MMACS Officer) any more than the other Shuttle systems. People were actually amazed at my self-restraint and willingness to let them do their job, instead of micromanaging them. The reality was, none of those people knew that, in my heart, I was really an IPS guy—and I was riding them for all I was worth! I'd written and tested the procedures, developed the rules and displays, and even been part of the development operations team for the system. I was sure no one could stump me on it.

What was really funny was that the training team didn't know my relationship with the system either—so I really looked better than I might have been.

Sure, I was an "expert," but I was an expert in the old ways. In actuality, I hadn't been a part of the IPS world since about 1986. I was simming with the system as Flight Director in 1994, and a lot had changed in the software over those years. Thankfully, the guys responsible for the system were old friends, and they were gentle with me when explaining that certain software had been upgraded, and how the procedures had been modernized from those first ones we wrote as junior engineers. It was, however, fun to show my expertise in at least something in the front room. I was still drinking from a firehose in many disciplines, and having the comfort of really knowing one of the major systems at the bit level gave me a place to stay grounded.

For me, one of the most amazing transformations in becoming a full-fledged Flight Director was learning to fly rendezvous. Finding another object in space and either grabbing it or docking with it is fundamental to any space program that wants to accomplish things in orbit. Mastering it was essential for us. Up until the time we were selected into the Flight Director training program, all rendezvous with the Shuttle had ended with station-keeping (stopping close to an object, and flying formation with it) or grappling the target using the Remote Manipulator System (RMS). But we knew that the days of docking to space stations were coming. In fact, I had been preparing the way for this in my work on the Russian docking system used by *Mir*—and slated to be used on the ISS.

Up until my selection as a Flight Director, I had been so involved with mechanical systems that in rendezvous simulations, the actual rendezvous itself was simply background noise in my headset. I understood the fundamentals and the orbital mechanics, of course—I had studied all of that when I first arrived as a co-op student. But the detailed procedures were someone else's responsibility. It was something I never had made the time to study. I do remember watching Flight Directors like Al Pennington running a rendezvous and wondering how a former

Instrumentation and Communications Officer (INCO) like Al could ever have learned so much about the details of rendezvous to make critical decisions on what seemed like very complicated scenarios.

Twenty years later, I had come to realize that I was the most experienced rendezvous Flight Director in the world—and I am still not sure exactly how I learned all of that stuff! You learn it one bite at a time. I know I could never write the mathematical equations or read the computer code for targeting software the way the experts in the discipline could do, but I know I could fly a rendezvous as a commander in the cockpit without referring to the twenty-page checklist; and I understand all the failures that could occur that would affect the successful completion of a docking. Even so, it wouldn't be as pretty as if I had grown up in the trajectory world. Once again, the old idea that I was the orchestra conductor and not a concert violinist comes to mind. I was no virtuoso, but I certainly knew how to run a team so that a magnificent symphony would come out of it.

One of the lasts things you have to think about when you finish up your training as a Flight Director and get ready for your first mission is your team name, or "color." As I said before, the tradition of Flight Director colors began when Chris Kraft, Gene Kranz, and John Hodge decided that they would be leaders of the Red, White, and Blue teams (respectively). The team colors were used as a morale-building device, but also for practical reasons—simulations and other activities for each team were scheduled by color rather than the name of the Flight Director. It worked, and it caught on, so the next couple of Flight Directors picked colors as well—Glynn Lunney picked Black and Cliff Charlesworth took on Green when they joined the first three on the center console in the MCC. The simple colors being taken, the next guys chose Maroon and then Gold—which opened up the idea of not only colors but minerals. Eventually, we had Flight Director colors for astronomical objects, gem stones, and constellations.

In the military, fighter pilots generally have nicknames or call signs. They are usually given by other pilots. You don't get to choose your call

sign—it's given to you, usually to remember something really dumb
you did in training. Ask a pilot how he got his call sign someday, and
you will often get a muffled response that tapers off to the inaudible. Or
they'll proudly tell you how they got it as part of a great story they tell
on themselves. But in the Flight Director world, you got to choose your
call sign—and it stays with you forever. Most new Flight Directors spend
more time than they are willing to admit picking just the right name for
their team—and I was no exception.

Colors didn't seem right, and neither did stars or constellations. But
I did have something I was very proud of, and my team name was suggested by one of my former Mechanical Systems teammates. He knew
that Minnesota was important to me, and he'd caught on that my ances-
try was from the northeastern corner of the state—the Range. The Iron
Range, that is. My grandfather was an iron miner at times—as were
most guys who grew up there, at least at times when they didn't have
other jobs. Ninety percent of all the iron ore mined in this country up
until recent times has come from the Minnesota Iron Range—an important factor in building a strong and powerful nation. And so I chose Iron
as my team name, and I proudly served the space program as the leader
of the Iron Team.

I announced my team name at the beginning of my first shift as a
Flight Director on an actual mission, and it was a proud moment when I
first used the call sign Iron Flight. The color was retired with me twenty
years later, when I left the agency—but the name stays with me no matter where I go.

My first mission assignment as a Flight Director was to STS-63,
a mission with a primary payload that I would have to go look up to
remember—but whose flight plan included the first rendezvous with the
Mir space station. I drew the assignment because of my Russian background, obviously. I was assigned to the planning team—the overnight
shift when the crew is asleep and nothing happens except replanning
the timeline. That was always where the new guys started out—it was
quieter and required less dynamic problem solving. The other Flight

Directors were the folks I had been working with for several years to get ready for the series of Russian missions, and so this assignment was a good place to break me in.

We weren't ready to dock with the *Mir*—we hadn't finished integrating the Russian docking system into the Orbiter—but we wanted to get something on the books with the *Mir* as soon as possible. Our plan was to fly a rendezvous and then fly around the *Mir* taking pictures and getting used to communicating between crews and control centers. One of the big worries we (and the Russians) had was what effect our thruster exhaust plumes would have on their solar arrays, and this mission would be our first chance to see what kind of a problem this might present. There were many, many hours of meetings and discussion over who would be responsible (and liable) should damage occur. We were happy with the idea of joint responsibility, but the Russians liked to say that "if everyone was responsible then no one is responsible." Interestingly enough, that made sense.

We had developed special Reaction Control System (RCS) jet configurations to allow us to fly in close without pointing jet plumes directly at the *Mir*. This involved using jets that weren't primarily pointed in the direction we needed them to thrust, but instead produced just a little thrust in that direction. It was inefficient but it worked. The problem was there were only a few jets that, used in pairs, could provide this capability. There was very little redundancy. Now remember that it was not uncommon for jets to develop a leak—some piece of crud might get in a valve seat, which then caused a slow seep of fuel or oxidizer. This was a problem because if it accumulated in the firing chamber, and then you went to fire the jet, you could get an off-nominal firing (oxidizer rich or fuel rich), which could damage the jet. In simple terms you might blow something up.

In any case the standard response to a leaking jet was to seal it off—and that meant isolating the other jets that lived on the same fuel manifold, since the only way to isolate it was to close that manifold. That meant a single leaking jet could take down three or four others. It was standard ops for a leaking jet, and we used the procedure many times on other Shuttle flights and in sims. But on this mission, the loss of the critical jets meant that we would lose the rendezvous, because without the special jet configurations to prevent pluming the *Mir*, we weren't going to be allowed to get close.

Well, of course, right after we got into orbit we had a leaking jet and had to close a manifold. We were still okay, but we couldn't tolerate any other failures. We were now zero fault tolerant to achieving the rendezvous. For this mission, the ascent team handed over to the planning team about five hours after launch. This coincided with the evening time in Houston. It was probably five in the afternoon when I came in to take my first ever shift as a Flight Director. The "usual crowd" was there. It was somewhat a standard thing for senior management to stop by the Control Center before going home for the day, so as I came in to begin my shift, the back row (behind the Flight Director console) was a brass-rich environment. I recall Gene Kranz was there along with Tommy Holloway—both senior managers, former Flight Directors above me in the chain of command. Randy Stone, head of MOD, was standing next to the Chief of the Flight Director Office (my immediate boss), Lee Briscoe. The office lead for Russian work, Gary Coen, was there, as was the Lead Flight Director for the Mission, Phil Engelauf. The presence of these folks didn't bother me; we all knew and respected each other. And although I was junior to them, I had proven I was ready, so it was just another day. No pressure. I conducted a standard handover briefing with my team and we took over control of the mission, dismissing the previous team who had put the crew to bed. No more than a half hour had elapsed from saying good night when I got a call from the front row... "Flight, PROP, it looks like we've got another leaker." Everyone stood up straight as I asked the usual questions—what jet, what were they seeing exactly, and could they be being fooled by instrumentation or something else. Sadly, this was a pretty straightforward and obvious leak—and, yup, it was going to close one of the manifolds that would make the proximity operations (getting close to the *Mir*) a No Go. We talked about it for a few minutes, and it became clear that either we closed the manifold or we continued leaking—and we weren't allowed to get close to the *Mir* with a leaking jet either. (It's not as if we were somehow allowed to just let it keep leaking by our own rules.)

Well, here I was. Years of practice and training had come down to this mission-critical decision on my very first shift. Fortunately, I thought, I have a peanut gallery of the finest ex–Flight Directors sitting right behind me. I figured it would at least be a comfort to confer with them before I made such a drastic decision—after all, this was not just a technical decision, it was a decision that had international political implications. So I turned around and saw all of these eyes look straight at me... silently. Finally, Randy Stone, head of Mission Operations, former Flight Director, and the man who took over Gene's job, looked straight at me and said in a firm voice over the airwaves, "Well, Dye, what the hell are you going to do now?"

I realized in that moment that I truly was the Flight Director. I had signed the logbook, I had signed for the crew and the vehicle. It was my decision, my responsibility, and the august panel arrayed behind me was not going to step in and take that away from me—or bail me out. I paused, smiled, and turned back to the front of the room. I keyed my microphone. "GC, Flight—let's arm the building for air-to-ground comm. INCO, make sure that we have a good link and let me know when that's established. PROP, I need your switch throws. And CAP-COM, let's get ready to wake them up—we have to close a manifold on the overhead panel."

Yup—for better or worse I was the Flight Director and it was now my job to lead.

Chapter 6

Plan, Train, Fly

Mission Operations—and Mission Control—have lived by a small number of mottos and slogans over the years. The most famous is probably Gene Kranz's "Tough and Competent"—a description he gave the men of Mission Control in a famous talk following the Apollo 1 fire (in which three astronauts died in a launch pad accident in 1967). He told all flight controllers to go back to their office and write the words "Tough and Competent" on their blackboards and to never erase them—they were now the creed for which Mission Control would be known. If you look carefully through the offices of Mission Control Center (MCC) today you can still see these words imbedded in faint outlines on old chalkboards—and written on many modern whiteboards by those who remember the past.

During the Shuttle era, a time of numerous missions and production processes that saw Mission Ops preparing and flying mission after mission, the phrase "Plan, Train, Fly" became the mantra of the organization—for that in a nutshell was what we did. We planned missions, trained for them, and then flew them. "Plan your flight and fly your plan" is an old aviator's motto that we took to heart and executed on a daily basis. For much of the Shuttle program, Mission Operations was comprised of roughly six thousand people—both contractors and NASA employees. Every one of them—engineers, managers, and staff support personnel—were dedicated to these three functions. Flight planning was a huge task that was about much more than defining a timeline; it involved trajectory design and analysis, cargo development and integration, and flight rules and procedures development. Shuttle missions might have looked the same to the outside world but each one was unique and took hundreds of thousands of man-hours to plan.

Plan Your Flight—Fly Your Plan

Of course, planning a Shuttle mission involved far more organizations than just Mission Operations. The way NASA was organized, the Space Shuttle Program Office was the top authority with what was done with the vehicles and the missions. They effectively "hired" other organizations within the agency to conduct these missions, with their oversight. For instance, the Kennedy Space Center (KSC) was responsible for preparing the vehicles for launch and processing the payloads, while Mission Operations at the Johnson Space Center (JSC) performed the detailed mission planning, training, and operations. Various mission managers within the Program Office itself managed each flight to make sure that everything came together to make the mission happen—and by everything I mean tens of thousands of little details. They needed to make sure that not only was the major payload and its team ready on launch day, they also had to make sure that the tortillas in the food locker were fresh and packed on time. Like I said—details.

A good rule of thumb for mission planning is that a flight was defined about a year to a year and a half in advance. Generally speaking, the concept of a mission was put down far in advance of that point so that the system knew what was coming. For example, the International Space Station assembly process took ten years, and the assembly sequence was mostly laid out before that began. But detailed flight planning, crew and vehicle assignments, and the actual generation of plans and products started in the year-and-a-half time frame. That is about when we would assign a Lead Flight Director and managers within the various elements of the Mission Operations Directorate (MOD) to herd the planning through the process. A lot of work happened up front to make sure that the mission fit within the capabilities of the program—that we could lift the weight required, carry enough consumables to fly the mission length necessary, and that the payloads physically fit and could interface with the vehicle.

Timeline and trajectory options were numerous at this point—did you need to launch at dawn or at dusk, or somewhere near midnight or noon? What would this do to the lighting for the deploy of a satellite? Where would the sun be relative to the pilot's eyes during a docking? Did the payload have a need to be over particular points of the earth at particular times in the mission based on their own event schedules? What kind of altitudes were required? These and hundreds (or thousands) of other questions were asked and answered in order to narrow down the options and pick an overall mission plan. It was an iterative process—oftentimes you had to throw away work when you realized you'd headed down a dead end. When that happened, we had to back up and start over.

Most of this work was monitored regularly by the Lead Flight Director as the mission developed. The Lead Flight Director would let the teams do the work and provide leadership to make sure that everyone was marching in more or less the same direction. At any one time a Flight Director might have two missions under development that they were leading, and they would also be assigned to several other missions that were being led by other Flight Directors. This process kept the half-dozen active Lead Flight Directors within the office busy and hopping throughout most of the Shuttle program. Lead Flight Directors wanted to make sure that they stepped in at key points in the mission-planning process to help make key decisions, but they also wanted to let the teams have ownership and do the work. Micromanaging smart people rarely works out well—but sometimes, they could get their heads down and build up a lot of speed in a direction that might not be desired by the program. A smart Lead held lots of status meetings and checked up with the mission planners often enough to effect changes when they were easy to make.

Crew assignments were made when most flights became "real" and the mission was announced. That was also about the time that we tried to assign a Lead Flight Director. These assignments happened about the same time so that neither the Lead Flight Director nor the Shuttle commander had a head start on owning the process. Good Leads worked well with good commanders to make sure everyone was happy with where a mission was going, and the best ones always remembered that it was the program that we were flying for—not ourselves or our organizations. Although we all cared deeply about doing the right thing for the program, NASA, and the nation, within every mission we also had to recognize that overall policy was set above our pay grade. I used to tell flight controllers that unless program direction was illegal, immoral, or fattening, they should be able to live with the requirements that were being given to us. We all had responsibility for doing our individual jobs well, but the final responsibility ultimately rested with the program manager who had to answer to the NASA administrator.

Getting a large army of engineers, planners, and technicians marching in the same direction takes a lot of time and effort, but it's vital work because changing directions takes a lot more energy than getting them started correctly in the first place. Knowing this, the Lead Flight Director and the program's mission manager always put extra effort into the early stages of the project. You could then let the designers of the trajectory and vehicle loading efforts go on autopilot to, as we said, "turn the crank." The result would be products coming out of the planning pipeline.

One of the things that the program mission manager, the Lead Flight Director, and the Ascent/Entry Flight Director kept tabs on during planning was the Ascent Performance Margin (APM). Essentially, if you assume that you know exactly how much the Shuttle can lift to the desired orbit and you are loaded exactly to that number, you have zero APM. Zero is never a good number when you're talking about margin, of course—so you always want to maintain that as a positive number. The Ascent Flight Director wants it to be a *large* positive number because that means they have some wiggle room if things go wonky on launch day because it can be used to expand the launch window—the time you have to get off the pad on any given day. The Lead Flight Director is always looking for a large number that they can then convert to a small number by loading more consumables (propellant and cryo) so that they can have the best chance of executing the mission on orbit and maybe extending it, if needed. The program manager wants that number to be as close to zero as possible because that means they are carrying everything they possibly can. I have seen cases where they used it at the very last minute to load mementos for space-flight awareness awards. This is when the rule of "illegal, immoral, or fattening" comes into play. In cases like this you sometimes have to allow for the fattening… after all, it was the Program Office's vehicle.

Things really got cranking in MOD about nine to six months before the mission. By this time, the plan was pretty well set—the payload operations were well known and drafts of all of the flight specific procedures (and the flight plan) should be available to everyone. Reviews became intense and nit-picky as it became obvious that the flight was real and was going to happen. Simulations were still months away, but it was time for folks to start going through the plan with a fine-tooth comb, verify procedures in the simulator, and start seriously working on contingency plans for the cases where something might not work right or took longer than planned.

Mission Operations had developed a well-earned reputation for being able to make complex real-time decisions quickly, accurately—and correctly. But in actuality, most of these decisions aren't made in real time—they are made months (or years) before, long before the spacecraft was hurtling around the planet at 5 miles per second. The planning process was where these decisions were made by way of long working sessions and meetings where options were discussed, potential solutions to potential problems were hammered out, and decisions were made on what to do when "X" happens. There is a big difference between making a decision and executing a plan, and executing was what we wanted to do in real time. In real time then, all you had to do was apply the premade decision to the situation that was happening in front of you. This made it look like the decisions were being made in an instant. In truth we were merely executing a plan.

This is of course the ideal way of doing business—just execute premade decisions when bad things happen. This was the way it worked much of the time. But the reality was that things happened that no one thought of in advance, and so there were real-time issues that had to be solved and decisions that had to be made. The fact that you have worked through the paths to take for *many* possible contingencies, however, leaves you with more time to handle the unexpected when it inevitably comes up. If you waited until the actual flight to work out all the potential problems, you'd be overwhelmed very quickly.

We all now remember the Apollo 13 mission, long before the Shuttle came along. This mission is still cited as the benchmark for how Mission Control operates and what it can do. There were so many problems to solve to safely bring home the crew of the crippled Apollo 13 spacecraft that most flight controllers didn't go home for three days. There were work-arounds for damaged hardware to be developed, new ways of flying the spacecraft to be explored, and lots of number crunching to be done on the trajectory and the available consumables. When the crew made it home safely, it looked like they had pulled the proverbial rabbit out of the hat. This is because they did. But they didn't do it from nothing. Flight controllers had been thinking about far-out contingency cases and failures for years, and while much of this work was done unofficially and kept in various "hip pockets," all that time and work proved useful when the oxygen tank exploded and the "far-out" cases became real. I learned from my mentors, the folks who were actually there, that virtually everything that was done to bring Apollo 13 home had been thought of before the flight and was in someone's head as a possibility before it was needed. They didn't have to make it all up in real time; they had at least gone through the thought process before, which gave them a huge head start on solving the problem.

Planning for contingencies was therefore a huge part of the preflight preparation process, and it probably took more time than developing the primary plan for each flight. It was what Mission Control and Mission Operations did. It was their bread and butter, the secret to their success. You don't have to be the fastest to get to the finish line first, sometimes it is just a matter of starting out earlier than everyone else.

Of course, a lot of the contingencies that were looked at before each flight weren't the work of an individual or a group of flight controllers brainstorming what might go wrong. Many of the best contingency cases (and solutions) came from another venue, another department. Enter the world of flight specific mission training and simulations...

Train the Way You Fly—Fly the Way You Train

In theory (and mostly in practice), flight controllers who were assigned to a particular mission had first to be certified in their position through the normal process of studying, training, and generic sims. That meant that when we scheduled simulations for a particular mission, everyone was already good at what they did—we didn't need to test them on routine operations or their judgment on handling cases that were common to all missions. What we needed to do was to rehearse the mission we were about to fly, or at least the parts of the timeline that were unique to that flight. We also needed to test the unique procedures and rules we developed for the mission. Our goal was to determine if we had actually thought of all the ways that things could go to worms, and discover the holes in our contingency planning.

In the early days of Shuttle operations, it was not uncommon to conduct two hundred hours of integrated simulations (which involved both the crew and the MCC) for a specific flight. This volume decreased over the years for a number of reasons. The first was budget pressure. It costs a lot of money to bring up an entire operations team, multiple buildings, communication links, and all the support facilities to run a sim. The second was that as we gained experience we had a better handle on the job of flying the Shuttle, and our deeper experience told us what we needed to train for and what we probably didn't have to worry about. The third was that the simulators and MCC got better. Early mission simulations often collapsed because the simulator wouldn't run right, or the communication network wasn't working, or the software didn't accurately predict the actual way the systems operated (or performed). Oftentimes, a full-up sim would grind to a halt as the sim team worked to troubleshoot the machinery we needed to proceed and, eventually, the session would be canceled.

One of the fun parts of early missions were the integrated long sims. These rehearsals were up to two days in length, and everyone ran in real time and worked shifts just like an actual flight. They were fun because we were truly immersed in a flight environment, and at times it was hard to tell the difference between a simulation and the real thing. In the middle years of the Shuttle program, we decided that these long sims took up too many resources that were needed for the multiple missions that were, by then, flying. So we cut them back and only used the technique for special cases of very new, very complex missions. By the end of the program, long sims were a thing of the past. We flew often enough that new folks could learn the techniques of handovers and round the clock operations by helping on real missions. It was a bit of a shame though. Many of those long sims truly showed us what we were lacking in our plans, and it was hard to develop really informative and interesting mission-unique sim cases that could be executed in just eight hours. Forgoing the long sims never truly harmed us in those later years, but they were experiences that the younger generation missed.

In developing a plan for mission-unique simulations, the Lead Flight Director and the Lead Simulation Supervisor (Sim Sup) would sit down and look for the key points in the mission that they felt could hang up the teams. Those would be the focal point for the orbit simulations for the primary teams. They would then pull out a fairly standard template for the ascent and entry sims, and then throw in a couple of sims for the nonprime teams (the second and third shifts that were not working with the Lead Flight Director). Once we entered the ISS construction era, the sims became pretty cookie-cutter. The core flight controllers sort of went into idle mode once the Shuttle was docked to the ISS. The Remote Manipulator System (RMS) and EVA (Extravehicular Activity) operators were just getting started at that point, of course, but then later in the assembly sequence even those tasks shifted over to the ISS side, which meant the ISS Flight Director and ISS team ran the operations and the Shuttle team mostly cheered them on.

Mission-specific simulations were not supposed to be used just to keep Shuttle flight controllers current. There were times when there was simply nothing for a Data Processing Systems (DPS) or Guidance, Navigation,

I was privileged to work my years as a flight controller, and my first several years as Flight Director, in the old Mission Control that was used for Apollo. We were encouraged by the spirits of those that had gone before.

We always tried to encourage our flight controllers to remember the missions of the past—in this photo, our on-console team celebrated the Apollo flight control teams on the fortieth anniversary of the first moon landing.

The old Mission Control consoles and systems were used until the mid-1990s to control Shuttle missions. The consoles featured black-and-white monitors and hardwired event lights. It actually matched the 1970s-era Shuttle design well, and improvements to MCC came along with upgrades to the vehicle technology.

The NASA Flight Director Insignia used only by those who have been responsible for a crewed spacecraft in flight.

Being the Flight Director meant keeping up with a huge amount of information from the vehicle, flight controllers, the crew, and the management team. Every shift was packed and required a lot of engagement.

The Spacelab Instrument Pointing System on STS-51F—my first major system responsibility in Mission Control. The rectangular box on the right with the protruding tubes is the Optical Sensor Package, which tracked stars to keep the system stable.

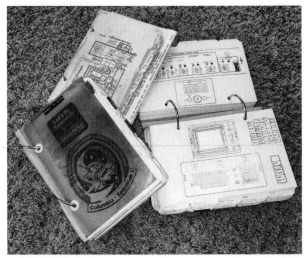

The "goodie books" I built during my early years as a flight controller served as shortcuts to the many thousands of pages of documentation and schematics for the systems I watched over.

Left to right: myself, Bryan Austin, and John Shannon (the Flight Director class of '93) during crew escape activities in our year of Flight Director training. It was the intent of such training to make us familiar with just what our crews would be going through in flight—including the limitations of wearing the bulky orange launch and entry suits.

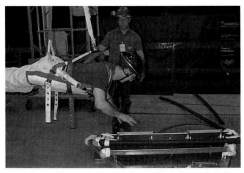

Here I am experiencing what it is like to work on a spacewalking zero-G simulation rig used for astronaut training. This rig allowed us to understand just how every action created an equal and opposite reaction—so if you turned a wrench on a bolt, you moved instead of the bolt.

You can only live on pizza and cookies for so long on console—sooner or later, you need a healthy snack. Without the ability to leave the Control Center for meals, we tried to have reasonable food during missions whenever we could.

The Flight Director job included doing press conferences during missions, like this one in 2009. We were sent to a weeklong "charm school" during our year of training to better understand how to work with the press. This training drummed the phrase "no comment" out of our repertoire.

It took a great team to fly a Shuttle mission, and one of the most important relationships for a Flight Director was working with his or her CAPCOM. Here I shared the last Space Shuttle Mission with my frequent CAPCOM, Shannon Lucid, one of our most experienced female astronauts, whom I worked with for many years.

A large group of current and former Flight Directors gathered in April 2011 to help dedicate the Mission Control Center to Christopher C. Kraft (front row, center)—the founder of MCC and the very first Flight Director. Kraft was a mentor to every Flight Director who ever held the position.

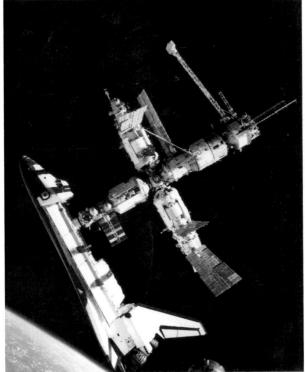

I was proud to see the Shuttle docked to the *Mir* Space Station on STS-71, and to be a Flight Director on that mission. It was a culmination of years of preparation, and the beginning of a lasting relationship between the United States and Russia in space exploration.

STS-71 Shuttle Commander "Hoot" Gibson shakes hands across the hatch with *Mir* Commander Vladimir Dezhurov during the STS-71, the first docking mission. It took years of work by thousands of dedicated space workers to reach this moment—and US/Russian relations were improved because of it.

The SRTM mast canister was mounted in the payload bay, and the mast extended out over the port side of the Orbiter. The radar mapper created an accurate and detailed topographical map of the earth that will serve as a baseline for all mapping for the next century—a satisfying mission for everyone involved.

I took some informal time during the last Shuttle mission to talk with my young flight controllers about the early days of the program. It was inspiring seeing such talent still with us that would continue to pursue human spaceflight well into the future.

GUIDANCE

It's sort of an honor to be memorialized on the cover of an Execute Package—even when they are poking fun at you. In this case, the young flight control team had a laugh at the length of my service as a Flight Director.

Popular Space

The What's New in Space Magazine --- *Special Paul Dye Edition*

The New Face of Mission Control
New Flight Director Spawns Hairstyle Craze

Dye Vows to Put Man on Mars by End of 70's!

Dye Explains How to Extinguish Fire in 43 EZ Steps.

Brady Bunch to Visit MSC!

Build Your Own Spacecraft Using Vacuum Cleaners!

NASA Discovers New Planet, Names it "Pluto"

25¢

Flight Directors, controllers, and instructors often served as surrogate crewmembers during simulations. Here I was the commander on the final Shuttle generic simulation of the program, an eight-hour training session that kept our crew—and Mission Control—hopping from start to finish.

During STS-131, I took time to go over the Shuttle Rendezvous Checklist and compare notes with Apollo 13 Commander Jim Lovell. Lovell helped develop rendezvous techniques for Gemini and Apollo that made the moon landings possible, and it was an honor to catch him up on how we used his work in the Shuttle program.

The International Space Station viewed from the Shuttle during STS-132. The last ten years of the Shuttle program were dedicated to building and supporting this million-pound laboratory, a crowning achievement for everyone involved in human spaceflight.

Execute Package humor was often aimed at Flight Directors. Flight Directors rarely do anything themselves—rather, they direct others to do things. Sometimes, those other things aren't even human!

and Control (GNC) Officer to do during a sim when the ISS was running an EVA along with cargo being pulled from our bay. Sure, the sim team could kill a computer in the DPS world or a black box in the GNC domain, but because the emphasis was on the payload operation it was generally just a matter of "if A fails, switch to B"—take two aspirin and call me in the morning. It could get slow in other words. That was when I encouraged good front room operators to let their back rooms watch over things so that they could pay attention to the big picture. We were always looking for new talent with the potential to be Flight Directors—and the ones who were willing to look outside of their discipline to spend time watching the big picture were the ones we wanted to find.

Training in the middle of the Shuttle program—back when we had significantly different payloads and weren't just going up to assemble the ISS—was more interesting. We spent a lot of time learning to work with our payload partners and they learned a lot about what it was like to work with the Shuttle team. Large science payloads generally had a big team of supporting scientists. They worked around the clock from their own control center or in a room within our Mission Control Center that had been equipped to support them. These Payload Operations Control Centers (POCCs) were run by the Payload Operations Director (POD), the Flight Director's counterpart and single point of contact for overall decision-making. I started working directly with PODs and POCCs in my early Spacelab days as a flight controller, and that experience gave me a leg up on many Flight Directors who grew up in a single, traditional Shuttle discipline that had little interface with payloads.

The Spacelab systems that I was responsible for as a flight controller were not Payloads but not exactly Shuttle—they were situated between the two. In essence, we provided Shuttle support to the experimenters. In that role I had to understand and work with the payload world at a level that most flight controllers did not. It was then that I learned just how important a simple little experiment could be and just how critical it was that we gave every one of them the maximum level of support so that they succeeded. It was easy to see how a simple container of fish eggs could be ignored in the big scheme of things—but this shortsightedness could ruin years of research in some scientist's field. I always felt that if they had gone through the effort to get on board the Shuttle—a daunting bureaucratic mountain to climb if ever there was—then we owed them the best ride and the most attention we could give them. So I welcomed sims that tested all these little guys as well as the major payloads that were more visible. It was pleasing to help an experimenter succeed—and to show them that they would receive the same level of ingenuity that our regular flight controllers gave when it came to troubleshooting.

At the end of the Shuttle-Mir Program we flew the first Alpha Magnetic Spectrometer (AMS) payload into space. This was a large device designed to look for antimatter particles (among other things). It was actually the prototype for a larger, more capable device that would eventually be placed permanently aboard the ISS. As was typical of a major science experiment like this, before we flew I spent a considerable amount of time and effort getting to know the experiment, the science, and the scientists involved with the project. The head of the project was Dr. Sam Ting, a Nobel Prize winner and giant of particle physics. His team was spread around the world, as is true with most major science these days, but was headquartered in Geneva. The AMS itself was being built in Zurich. When we were just beginning the detailed mission-planning process, I took a quick trip to Switzerland to get to know the team and see the hardware.

My most vivid memory of the trip was of sitting down with Dr. Ting in his office and he asking me what I knew about particle physics. I was sad to report that the little I had learned back in freshman physics at the university had long been overwritten in my brain by more current and pressing matters. I admitted that what I did know these days was probably from educational television. Without missing a beat, he said, "Well then, we need to get you up to speed." He then picked up a piece of chalk and began giving me a basic lecture that ended a couple hours later. In that time he taught me a good deal about just how the experiment worked, what it was looking for, and how this was going to benefit our overall understanding of the universe. There's nothing like learning your particle physics from a Nobel Laureate in the field. It was kind of him to take on a simple engineer as a student, but it created a bond between us that kept us working together for the rest of the mission and let each of us know that we were committed to doing the best we could for the common good. Being a Lead Flight Director for a science mission was like that—it gave you great opportunities to step outside your own field and learn something about the contributions that the Shuttle was making to science.

It turned out that the sims for the AMS experiment were interesting. It was a huge system, and keeping a group of scientists all headed in the same direction while the clock is ticking and you're going around the planet once every ninety minutes is, indeed, a challenge. By the time they recognized a problem and made everyone else aware of it, the time to activate the experiment oftentimes had already passed and the danger was in missing the *next* opportunity to take data. We often encouraged POCC teams to sim on their own, and Flight Directors would go to their facility to participate (as if a full MCC was involved)—before we got to the full-up sims that cost a lot and were limited in number.

It was good that the AMS guys did stand-alone training and all the integrated sims, because when we got that mission into orbit, we discovered a failure in the high-rate data system that was designed to give them real-time insight (on the ground) of the data that was being collected. After determining that we just weren't going to be able to fix it, we recorded the data on board and then dumped small pieces of the data through our standard-rate telemetry system once every so often. This gave them glimpses of what was going on and allowed them to make system adjustments. It was like trying to watch a baseball game through a keyhole ... you miss a lot of the action. They did remarkably well, however, because they took the training time seriously.

Of course, our own in-house training wasn't all done in integrated sims. Shuttle crews did a considerable amount of stand-alone training before they even got into the integrated training, and most of the time they welcomed observers and participants from the flight control teams to these training sessions. This was especially true for the specialized flight controllers where the trainers and flight controllers were all the same people. The EVA, RMS, and rendezvous disciplines were like this, with the same people training the crew and then serving in the Control Center during the mission. Occasionally Flight Directors would drop in on these sessions. That always made for fun. Plus it was far better than sitting through a meeting or working in the office.

Because I personally had a significant background in underwater work (I worked my way through college as a diving instructor and technical diver), I was certified early on to dive in the Neutral Buoyancy Lab (NBL) and the smaller water tank that preceded it, the Weightless Environment Training Facility (WETF). As a flight controller, I helped develop some of the EVA procedures related to the Spacelab IPS because I was familiar with the hardware. This was in the old WETF, which really was just about the size of the Shuttle payload bay. When I became a Flight Director, I could go over to the WETF control room when I had a crew in training and watch the progress of the EVA on the TV monitors. The EVAs were confined to a small space (the payload bay), and so it was easy to keep track of what was happening and how close (or far) the crew was from the airlock, the tools, or the payload. I'd usually also take a quick dive to observe the general characteristics of the EVA tasks, and then just watch from the surface on subsequent days.

But when the ISS came along, we moved to the new facility—the NBL. It was the size of a football field and 40 feet deep. The ISS itself was (and is) immense, spreading over that proverbial football field. While they didn't have the entire thing in the pool, they could simulate as much of it as we needed. It was so massive that it was easy to lose track of just where crewmembers were working. Watching the feeds from the helmet cams and the occasional view from a TV camera could just make the landscape more confusing. This was because the crew's helmet cams were close to the structure and didn't reveal where they were in relation to much anything else. So as a Flight Director, I often liked to jump in the water with a crew as they went through the EVA for a mission just to make sure I had a good feel for the whole process. It was a nice way to spend six hours, floating in a crystal-clear pool as warm as bath water and not having to answer the phone.

We also took the chance to fly with our crewmembers in the simulators when they were trying a new rendezvous or docking technique. We had a mission to *Mir* early on in the program when the *Mir* lost several channels of its attitude control system just before our launch. This meant that if they had another failure, the *Mir* might well be drifting in some random attitude when we got there, meaning we wouldn't be able to use our standard docking techniques. We sent the commander, Jim Wetherbee, over to the Shuttle Engineering Simulator. This simulator had a dome-shaped visual system; it gave you the best view of 3-D activities, which were necessary to try and dock with a slowly tumbling target spacecraft. We had incredibly small tolerances for errors in docking alignment (on the order of 3 inches) and rates, and it was challenging as heck for Jim to figure out where the *Mir* was going and how to match the Shuttle's rates and attitudes with it.

I made it a point to spend a lot of time in the dome while this issue was being worked out—after all, the mission was planned, everyone was trained, and this was the long pole in our tent regarding a Go or No Go for launch. These sessions went on for a couple days until we felt we had a workable procedure and were comfortable with launching. It was fun to help the guys develop a new way of flying and watching as we looked for pitfalls and uncertainty in what might happen—and meanwhile develop new criteria for when we could dock and when we had to knock it off. In the end, we were comfortable that we had a plan that would work. It turned out that we didn't need any of it—the *Mir* was stable when we got there and the docking went just fine. But the training process, once again, gave us a deeper knowledge of our available options and operational redundancy.

Training was also used as a method of certifying the flight procedures and rules that were specific to a mission. Although most of the generic procedures were carried from flight to flight with only minor updates (usually related to changes in hardware or software), we always had new and unique procedures for the specific mission, and these had to be proven in some way or another. The sign-off for noncritical procedures often happened during tabletop reviews. That is, the crew, controllers, engineers, and trainers sat down in a conference room and walked through the procedures line by line to make sure that everything was correct. If everyone was happy, then they were signed off. Small errors were often caught by someone, which made the time-consuming and tedious process worthwhile.

Fly the Plan

Planning a mission took a large number of organizations. It really took parts of the entire agency. Training for the mission, however, was mostly the province of the MCC teams and the crew (and all the associated training organizations that were part of Mission Operations). But once it came time to fly, the entire NASA community once again got involved—flying a mission was just a huge endeavor.

It was necessary, of course, to have a focal point for flight, and that point was the Mission Management Team (MMT). The MMT consisted of the senior managers of all the organizations involved in the flight, and it was chaired by the Shuttle Program Manager (or more accurately their deputy for operations). The MMT could number up to two dozen members at times, because flying a Shuttle affected a lot of people. Mission Operations, Engineering, the science community, the launch teams at KSC—these were the big ones and the most obvious. But there were also payload representatives, the manager of the ISS program (when we were flying to the Station), the EVA Program Office, and other organizations such as NASA Safety and Mission Assurance. While the flight control team—or more accurately, the on-console Flight Director—had real-time authority to make any decision they felt necessary for the safety of the crew and the success of the mission (as defined in the flight rules), the MMT had overall authority for non-real-time direction of the mission.

Essentially, any major decision that was not already covered in the pre-mission flight rules (and couldn't wait until a meeting of the MMT was convened) was the prerogative of the Flight Director. This is what we were paid to do (as well as a thousand other things, of course).

Familiarity with all the decisions that went into planning the mission was what gave us the ability to make those decisions, if necessary. If, however, a situation arose that wasn't already talked about in the pre-mission time frame and didn't have a premade decision to cover it—and there was time to get the MMT to weigh in—we tried to do that. After all, the Flight Director didn't actually "own" the mission—the program did, and it was only right for the MMT to make critical decisions that might change the direction of the flight.

This all sounds nice and clear-cut, but in reality it wasn't. If a situation arose that demanded a complete change in direction, it was the job of the flight control team to lay out the various options, determine the implications of each possible course of action, and make a recommendation to the MMT on what they thought was the best way to go. Oftentimes, it took quite a bit of education, and answering lots of questions, to get the MMT headed in the proper direction. The good news, of course, is that the MMT was generally composed of people who once worked in the trenches—as flight controllers, Flight Directors, engineering systems managers, mission scientists, and the like. So rarely were we starting from scratch when it came to bringing the MMT up to speed. The MMT was also the body of folks that were briefed before the mission (or at least they were supposed to be the ones who were briefed) and who approved the plans for the flight—both nominal and off-nominal.

For much of the Shuttle program, the head of the MMT was a former Flight Director. This meant the MMT was led by someone who knew how to wrangle a decision out of a group of technical people. At times the MMT would be led by someone who did not have a lot of opera-tional background, and that could really slow things down as they were brought up to speed so that they could understand the implications of a decision. More often than not, however, the MMT followed the recom-mendations of the experts from the various organizations, such as MOD and the flight control team.

Once the decision was made to launch, KSC had the responsibility for the vehicle—and all operational decisions—until T-0 (Time Zero), when the explosive bolts fired and the vehicle lifted off. At that point, control shifted to Houston, and MCC took over. The ascent phase of flight was an eight-and-a-half-minute ride of high energy that required a huge number of decisions, all of them made with no time to consult anyone outside the team. The flight controllers had to know their stuff, know their nominal and off-nominal procedures, and be able to recognize failures with enough time to take action. Coming off the launch pad, you had a number of different options. If everything went fine, you could make it into your nominal planned orbit. If you had problems with an engine (or engines), or anything that prevented reaching the planned orbit, you could hopefully still limp into a safe low orbit—this was known as an ATO—Abort to Orbit.

If you had a problem with engines or systems that required you to get back on the ground as soon as possible and you were still not going too fast, the Shuttle could be turned around in a complex procedure that was called the Return to Launch Site Abort (RTLS). John Young, the famed astronaut from the Apollo era, used to refer to the successful execution of an RTLS as, "Ten consecutive miracles, followed by an act of God." The RTLS involved turning the vehicle around backward, flying into the exhaust plume as you slowed it down, and then, after coming to a "stop" about 300 miles offshore, continuing to thrust so as to build up velocity for a ride back toward the Cape. During this time, the Shuttle is lofted up to about 400,000 feet in altitude, the fuel supply is emptied, and the External Tank is dumped before establishing a gliding flight that hopefully has enough energy to make it back to the KSC runway. It was jaw dropping when described, and we executed it tens of thousands (more like hundreds of thousands) of times in simulations over the life of the flight program. But we never flew one. For a pilot, it was a challenge to fly in the simulator with a good crewmember (or two) to help, and it was quite exciting to try on your own. Letting the computer fly most (or all) of the tricky part helped.

If you had a problem that prevented you from making it around Earth at least once, yet you had gotten up enough speed such that you were no longer able to make it back to the Cape, then you could fly a Transatlantic

Abort (TAL). This took you across the Atlantic to a landing site in Spain, France, Morocco, or some other runway in Eastern Africa. Executing this maneuver would take the Shuttle across the Atlantic in twenty-five minutes or less, an astonishing feat to fans of early aviators like Charles Lindbergh.

Finally, if you had too much energy to get slowed down for a TAL, but not enough to make it to an ATO, then you could do an AOA—an Abort Once Around. This took you around the planet in ninety minutes to land at White Sands, New Mexico. It was less challenging procedurally than either an RTLS or TAL, but it exposed you to systems failures (such as a cabin leak or loss of cooling) for a longer period of time.

The most interesting thing to me as an ascent flight controller was that as the ascent went on and your abort mode options dropped away, life got simpler because you had fewer branches of a decision tree to go down. This was true not only as you passed abort points, but it was just as true as you experienced system failures. The more failures you had, the fewer options you had left until you reached the point where if anything else failed you were out of luck. At that point, there were no more decisions to be made. You just had to cross your fingers and watch from that point on.

When the Shuttle made it into orbit, the crew and ascent teams had a couple hours' worth of procedures to perform in order to change the Orbiter from a rocket ship into a space station. These tasks effectively activated the systems that allowed the Shuttle to operate and work in space. This involved turning off a lot of equipment that would not be needed again (such as the main engines), or that wouldn't be needed again until entry (such as the hydraulics), as well as activating things that were necessary to stay in space. The most important of these tasks was opening the payload bay doors so that the radiators could see space and provide cooling. During ascent, the vehicle used the flash evaporators to keep the systems from overheating, but the water supply for those wouldn't last very long if you couldn't get the radiators working—so that was high priority.

After cooling, there were computers to reconfigure (we didn't need all five General Purpose Computers [GPCs] running ascent software anymore—one for Guidance, Navigation, and Control [GNC] and one for Systems Management [SM] was sufficient until you got into rendez-vous or entered into another flight-critical phase). You had to get the crew out of their suits, stow the seats, and unstow all the orbit equipment. The Remote Manipulator System (RMS) needed to be checked out if it was going to be used (as it always was after the *Columbia* accident), and the communications systems had to be deployed to the orbit configuration. This time period was called post-insertion and there was an entire check-list dedicated to it. When the crew reached the point where we knew that we could sustain the vehicle in orbit at least until the next day, the crew was given the "Go for Orbit Ops" and everyone could relax a little bit.

The typical Shuttle mission in the later years lasted about ten days, whereas earlier missions averaged about a week. Regardless, the time-lines were always packed to get the most out of them, and the crew and ground teams quickly settled into a routine. The MCC worked twenty-four hours a day, of course, and three shifts were required to operate a normal mission. Each shift worked for eight hours, plus an extra hour for handing over to the next team. It rarely worked out this cleanly, of course—the challenge was that we always tried to keep the ground teams synced with the crew. The vagaries of orbital mechanics and sun-rise/sunset times at launch and landing sites meant that we were usually shifting the crew by an hour or a half hour a day, meaning we had to shift the ground teams as well. The result was that sometimes you got a string of eight- or ten-hour shifts. It wasn't until the ISS era (with its rou-tine sync to Greenwich Mean Time) when flight control teams settled into a reliable shift schedule that never changed.

It was customary for the Lead Flight Director to take whichever shift had the key elements of the mission occurring during that time. More often than not, this was the crew's morning—what we called the Orbit 1 shift. But it was not unheard of for some major events (such as satellite deploy) to be scheduled in the crew's afternoon and, in that case, the Lead Flight Director might take Orbit 2 for the mission. The one constant was the planning shift, which occurred while the crew was asleep—except, of course, on those missions where we split the crew into two shifts and ran a twenty-four-hour-a-day operation on orbit. In such cases there was little difference between the three MCC shifts. The routine (if you could ever call any Shuttle mission routine) was to execute the plan you came up with preflight to the best of your abilities, recognize failures or hiccups to the plan as they occurred, and then replan the next day (or the entire mission) in an effort to zero in on accomplishing the mission goals. The crew and all the teams worked together to make that happen.

In aviation (and spaceflight), the old saying that you should "plan your flight and fly your plan" has been proven to be correct and necessary time and time again. Last-minute changes that are not driven by the need to address failures usually end up causing problems—or at least they cause more trouble than they are worth. Getting "creative" in the middle of a flight usually means that you're improvising a plan without taking the time to think about the implications of those changes—and that rarely works out for the better. While it is nice to take advantage of opportunities when they arise, those opportunities often come with traps and pitfalls that you don't expect.

One example is a story I was told when I first went to work for JSC as a co-op student. Several of the members of the section where I was assigned had recently been involved with the Skylab reentry in 1979. Skylab had been left in orbit by its last crew in the first half of the 1970s. The orbit had decayed a bit each year, as orbits always do. Atmospheric drag slows any orbiting object down, and slowing it down drops the altitude. A lower altitude has more drag, and eventually the object slows below orbital speed and enters the atmosphere. NASA's stated plan had been to have the Shuttle flying before the orbit got dangerously low, and to use the Shuttle to attach a booster to keep the space station from reentering. Unfortunately, the Shuttle was delayed, and it became apparent that nothing was going to save the Skylab. So a small team of flight controllers was organized to reactivate it and see what they could do about making the entry as controlled as possible.

It sounded like a science fiction episode. In today's terms they basically had to hack the old and degenerating systems to get enough power to run the computers and communication systems, and then bring up the gyros and sensors to see if they could control the attitude. Without enough fuel to actually fly the craft, all they could do was control the drag—make it greater or less, depending on the attitude. If you flew it in a streamlined attitude, you could keep it up longer by generating some lift. If you flew it in an attitude for maximum drag you could increase its rate of descent. If you tumbled it, you got essentially a ballistic reentry with a known trajectory.

The team figured out pretty well how to fly it this way, and when it came down to the endgame, their plan was to tumble the spacecraft and turn off the computers and telemetry once the trajectory was known. Their target would be a mid-ocean graveyard somewhere in the world. In the final weeks, it became apparent that a good disposal site was in the Indian Ocean, and they set up their plan to tumble the spacecraft over the continental United States and turn everything off. It was a plan that assured a good final watery resting place for any debris that survived all the way down to Earth's surface.

Unfortunately, at the last minute (according to the story I was told), a directive came from a high-level official, someone who had not been involved in the planning and, thus, did not know all the details and reasoning behind the scheme. This individual wanted the team to go ahead and tumble it, but to leave the telemetry on so that the breakup could be recorded. This had been thought of by the team, of course, but they had not had enough time to look at all the systems implications, all the ins and outs of the software to see what the spacecraft might do in that configuration. They objected to the change because of the old "plan your flight and fly your plan" rule. They were told to leave the telemetry system on.

True to form, fate stepped in. They tumbled the spacecraft as planned and turned off the guidance computer. Skylab went out of contact as it left the US, but when they next saw it from the Ascension Island tracking station in the middle of the South Atlantic, it was once again flying in the low drag, high lift attitude. Later analysis (after the reentry) showed that there was some computer code that looked for a case where the spacecraft was tumbling and because normally that was a "bad" thing, it reactivated the attitude control and set it to the last previous commanded configuration—which in this case was low drag.

The result of this last-minute change to the plan is that Skylab flew in low drag mode through the entry, which meant that it flew farther than predicted. As a result, instead of ending up in the middle of the Indian Ocean, it flew to the eastern edge of it...some pieces actually ended up hitting land in Australia. No one was hurt, but it was embarrassing— and all because someone in power, without operational experience, decided that the carefully thought-out plan should be changed with little thought. Plan the flight...then *fly the plan!*

Planning the mission, training for it, and then executing it—that is what we did for over three decades. It is important to note one additional thing could be added to this mantra—and that is to learn from what you did and then bend that learning back into the next planning process, the next training sequence, and the next flight. We never stopped learning. Most of the learning was captured in mission debriefs. Debriefs occurred all across the organization—from official crew debrief sessions to unofficial intradiscipline debriefs. Debriefs could take weeks to complete. Some might even occur a month later, after data had been retrieved and analyzed. The results of all these debriefs were captured in organizational documents that informed future missions and made the Shuttle flight program better every time.

The prevailing philosophy was that making a mistake in and of itself was not a big problem. The problem came if you failed to learn from that mistake or made the same mistake twice because you ignored it the first time. From mission to mission, or from day to day within a given mission, there were always things to be learned and efficiencies to be gained. Oftentimes, the lessons learned in planning or training weren't good enough to tell us what we really needed to know. This meant we had to climb the learning curve in short order as we went zinging around Earth every ninety minutes.

Chapter 7

Shuttle-Mir

It was a gray day, with heavy smoke hanging in the air as six Americans stared out the window of a small Nissan van when it left the Sheremetyevo airport outside Moscow, headed for the center of the city itself. I couldn't help but think just how improbable it would have all seemed just a few short years ago for me to be here—guests of the Russian government, in a place that I had grown up thinking of as a target for our country's nuclear forces. Moscow! Sure, NASA had worked with the Soviet Union on the Apollo-Soyuz Test Project in the mid-1970s—but nothing had really happened since then. Here we were, a decade and a half later, about to sit down with our counterparts in the Russian space program and figure out what we could do together in space. The gray buildings matched the gray skies, and the colorless cars and trucks, all looking like a dirty version of a European city. No smiles—I had noticed that already, and even though I had been prepared for it by paying attention in our cultural briefings, it was so noticeable for one who has lived his life in America. The thing is, Russians aren't sad all the time; they are much like us. Russians just considered it idiotic (literally) to walk around with a smile on your face for no particular reason. It was going to take some getting used to.

As mentioned previously, the Shuttle-Mir Program grew out of a need to work out a way of keeping the post-Soviet rocket engineers working on peaceful projects rather than letting them be hired by countries interested in developing nuclear missile technology. Once the decision was made to fly the Shuttle to the *Mir* space station, we had to figure out how to do it—and what we would do when we got there. The "how" was solved by pulling out a Russian docking system that would fit both their vehicle and ours (but had not had a chance to fly). The "what we would do" was solved when we created a series of missions to bring astronauts to the *Mir* for long stays. We eventually flew to the *Mir* ten times—one of those was a nondocking mission—and traded out a long series of crewmembers who lived and did science aboard the *Mir* over several years.

As important as the science performed by the crewmembers was, what we really gained was years of joint operational experience working with our Russian counterparts. This became much more important as the Shuttle-Mir Program morphed into Phase 1 of the new International Space Station (ISS). The original stand-alone program became a part of the ISS when the Space Station Freedom was redesigned into the ISS and brought in the Russians, Europeans, Canadians, and Japanese. It became clear that the way to a large station in space was to make it international, and to use both Russian and US components and modules to form the majority of that station. To make that work, we were going to have to develop long-lasting, detailed working relationships based on experience and trust.

When we first sat down across the table to talk with our Russian counterparts, it became apparent that while there were some differences, overall we found that our two programs were a good example of parallel development. Our goals when we sat down to fly a mission, for instance, were pretty much the same—accomplish the mission and bring our crewmembers back safely. Those of us raised in the days of the Cold War might have had some preconceived notions about the Russians' motives and methods of operation, but any thought of evil intent on their part evaporated when we got down to discussing the nuts and bolts of what we both did. The laws of physics and engineering are universal—not nationalistic. Trajectories worked because of physics, and so did the strength of materials, and how fluids and electronics worked. While the US systems tended toward electronics and computer sophistication that at the time had not yet been reached by the Russians, the solid and robust nature of their structures and systems served them well when it came to reliability.

The Shuttle program had never been classified, so we didn't have to worry about sharing what we knew from a legal standpoint. Sure, there was a natural reluctance on some folks' part—on both sides—to open up about things that we felt were our top developments. But we eventually shared most of the design data necessary to dock our spacecraft and share astronauts. Operationally, when one of our astronauts was on board a Russian spacecraft, they played by Russian rules, used Russian techniques, and spoke Russian. Likewise, in the other direction when they were on the Shuttle, they worked the way a Shuttle astronaut worked. Russians were responsible for anyone on their spacecraft, and we were responsible for anyone on ours.

While it was obvious that the astronauts and cosmonauts were going to have to spend a significant amount of time learning each other's languages in order to work on either side of the interface, we faced challenges of our own on the ground with communications as well. We started off with the assumption that both sides should be able to work in either language, so we signed up a group of us for Russian language lessons. It was pretty easy for us to observe that more Russians already spoke English as a second language than there were Americans who spoke Russian, so we felt we should catch up. Although most of us were very experienced and educated engineers, it proved to be quite a challenge to fit the language training in with all the other work we were doing. A few flight controllers learned a great deal of Russian, but most of us had to admit that we were going to have to rely on interpreters for years to come.

The interpreters we used in the US were mostly Russians who had immigrated to our country. They were generally highly educated people with advanced degrees and outstanding academic records, but not a lot of them had space industry backgrounds. They could speak both English and Russian very well, but the problem was that much of the time we didn't speak English—we spoke "NASA." Half the words we used were acronyms, a shorthand language that had developed because using entire phrases was just too cumbersome and took too long. If we had to say "Reaction Control System" every time instead of "RCS," we simply wouldn't have had enough hours in the day to finish a discussion.

So before we could really use these outstanding interpreters, we had to send them to our own language school to teach them a lot about space systems and trajectories. Fortunately, they were quick studies. Once they had picked up the jargon and the science behind what we did, they could then figure out how to translate what we were saying into Russian. Of course, this worked both ways, so the Russians had to teach the interpreters their own jargon, and those terms had to be translated into their equivalent English (or NASA-speak) terms. I am quite certain that if an outsider who spoke and understood both Russian and English had listened in on our conversations, they would have been just as confused as a layman listening to a pure NASA-speak exchange.

Because it took a fair amount of time to conduct business through a translator, we quickly figured out that having the Russian Flight Director talk directly to the American Flight Director in real time, while events were happening in space, just wasn't going to work. There was just too much overhead in the communication to execute real-time operations. So we created a new console position in the MCC (Mission Control Center), the Russian Interface Operator (RIO). Originally, the O stood for Officer, but some felt this sounded too military—so it went back and forth over the years, with either usage being acceptable at some point in time. The RIOs were senior flight controllers drawn from their own disciplines to serve temporary duty as RIOs. "Temporary" turned out to be for the duration of Shuttle-Mir for many of them—or at least until they were selected to be Flight Directors. Indeed, several of the early RIOs found it to be a good path to the Flight Director Office because it gave them such a good insight into the overall operation, as well as tremendous visibility among the Flight Directors already in the office.

A typical discussion between the Russian Flight Director in Moscow

and the American Flight Director in MCC began with the Russian calling their own version of the RIO, the PRP (which was a Russian acronym), who then called our RIO. The translators would be listening for
such calls and would translate for the RIO. The RIO would then call the
MCC Flight Director and pass along the Russian message. The American
Flight Director would then formulate an answer, pass it back to RIO, the
translators would turn it into Russian for the PRP, and the PRP would
pass on the response to the Russian Flight Director. You could sometimes go get coffee while waiting for this whole loop to come back with
the next response.

Of course, the RIO served more than just the Flight Director. They
were there to pass information and questions back and forth between
any MCC discipline and any Russian discipline, so they were very busy
and constantly in demand. Their loop was a constant barrage of English
and Russian, with conversations going on about trajectories, jet configurations, consumables, cabin pressures, and timelines. There were always
failures being discussed and plans to change as a result. Overall, the
coordination worked well because of the preparation and relationships
that were built before the missions, in countless hours of meetings, both
in Houston and in Moscow. These meetings began as soon as the joint
program was approved, and continued well past Shuttle-Mir. In fact, they
continue in some form to this day with the ISS.

We found that once we started putting crewmembers on the *Mir*, the
normal interaction we built from the MCC to the Moscow Mission Control
Center (TsUP) wasn't enough—we actually had to have around-the-clock
presence in Moscow, and the Russians needed the same thing in Houston.
A small control suite was set up for the partners in each control center, and
from the mid-1990s until now, there have always been American operations teams in Moscow, and Russian teams in Houston. We developed an
entire apartment building in Moscow to house our teams, modernized and
equipped to European standards; and the Russians had a constant presence
in a set of accommodations in the US. Our teams were in Moscow to support
the astronauts on the *Mir*, and the Ops Lead was a very experienced MOD

(Mission Operations Directorate) person assigned to be on the ground in Moscow. They were in charge of making sure that everything was done in a safe manner that resulted in a successful flight for their astronaut.

The first astronaut aboard *Mir*, Norm Thagard, flew to the space station on board a Russian Soyuz vehicle. He and his support team pioneered the joint operations concept so that by the time we sent the first Shuttle to dock with the *Mir*, drop off Shannon Lucid, and return with Norm, we had a good working relationship with the Russians. We could solve major problems quickly while effectively negotiating things that required less immediate action. We learned a few things during these early days, probably the most important being the Russian way of doing business.

We developed a saying: "*Nyet* [the Russian word for "no"] does not mean No...it means Not Yet." It was a tongue-in-cheek phrase, but it really turned out to be a truism. Any request that popped up from the American side that required the Russians to do something was usually met by a *nyet*. But it didn't really mean that they were saying no. It meant that they needed more information, that they wanted to think about it, and that they probably wanted to negotiate. Russians, we learned, are outstanding negotiators. Despite the fact that most of the Russians we worked with grew up in a Communist system, they are probably better capitalists than most Americans. They start from no, and then make you give them concessions to get to yes. It's a powerful technique, and one that we were rarely able to use ourselves with the Russians, because we were generally told to make things happen no matter what it cost us in terms of giving things away.

I do remember one time in which I beat their *nyet*. I was leading a team that was trying to set up a Y2K test between our system and theirs. For those who don't remember, Y2K was the big worry that when the clocks and calendars all rolled over from the year 1999 to the year 2000, there would be major computer outages and failures due to shortsighted programming. Although many predicted global chaos as the interconnected computer systems collapsed at the stroke of midnight, it turned out to be a pretty big yawn (as most who understood technology expected) when it happened. Nevertheless, we were directed to test the combined MCC-TsUP systems to make sure that the rollover occurred without incident. Conceptually, this was easy—just set the clocks ahead and watch it make the transition—the computers only knew the time to be what the humans told them it was.

Our team spent a week in TsUP negotiating the tests—it was going to cost money because it was going to take time. I let the folks who actually knew the details of what was going to be done really do the talking. I was there simply to make sure that the team was making progress. The Ground Control personnel worked well and diligently, and they came up with an agreement that was ready to sign by the Friday session—we were leaving on Saturday morning. The head of the Russian organization we were dealing with was well known as a tough negotiator. He was gruff and well connected (which in Russia generally means operating both over and under the table with government and quasigovernment organizations). As we were beginning to wrap things up, he took control and stated that, unfortunately, everything that we had talked about and agreed to was going to have to be reviewed because he felt that it was going to take too many resources and cost more than he had available. It's worth noting he was in charge of the ground facilities—we rarely had this type of difficulty with our in-flight counterparts.

It was clearly a play for more concessions on our part, or more money—or both. I had been on the receiving end of this kind of thing often enough that I was tired of it. I asked the interpreter for the floor, and I basically said that it was too bad we couldn't come to an agreement, that we simply were not authorized to give any more resources for the tests, and that when we got back home I would have to report to my superiors that we were unable to sign a protocol. I bluffed a little, saying that I would report this back to the NASA administrator directly and he would let the White House know. Now, I knew that my Russian counterpart was under pressure to make the tests happen, and that failure of the talks were really not an option for him. I was calling his bluff—and frankly, I really was ready to walk out. It wasn't going to cost me any-thing, because we frequently came up empty. Losing a point or simply tying was about the same on the scorecard.

I told my team to pack up their things and said again to my Russian counterpart that I was sorry, we would be back in touch, and our superi-ors would have to give us more direction. Now I knew that he considered himself to be the ultimate power in his fiefdom and that he didn't want to have to admit that he could be ordered to change his mind or give conces-sions by anyone. As we were walking out the door, I heard him address me directly, in English (the first time he'd spoken in English all week). "Paul, Paul...please come back, we can sign, we can sign...I think this is important, and I will find the money in my budget somewhere." He'd blinked, and I can't say I felt anything but good about it. It was not often that we "won" in this game—not that we really lost, it's just that often we felt that way. In any event, I was glad to feel like I hadn't let them push my team around.

It was on a different trip to Moscow when things were really put to the test, and it revealed how the Russians reacted to a true emergency. I was in Moscow for just a few days, negotiating the final form of the flight rules for an upcoming Shuttle visit to *Mir*, for which I was the Lead Flight Director. Mike Foale was the astronaut on *Mir* at the time, and things were going fairly well—well enough that while I visited with the Ameri-can team and the Ops Lead supporting Mike, I really wasn't plugged in to the day-to-day operations of what they were doing.

At one point in our small group meeting, Victor Blagov, the overall Lead Flight Director for the *Mir* (he was essentially the equivalent of the chief of our Flight Director Office) said that we should take a break and go into their control center. They were about to dock a Progress supply ship to the *Mir*, and he wanted to be there for the procedure, which was going to be done manually instead of using the usual automatic system. The Progress vehicle was essentially an unmanned Soyuz used to haul cargo up, and then burn up trash on reentry. Since it was unmanned it had no manual controls, but the Russians were working on a "radio control" mode where the crew in *Mir* could fly and dock it remotely. He suggested that I watch from the balcony with my interpreter. We settled down just as they were getting deeply into the procedure, and it quickly became apparent that something was not going right. There was a cluster of people standing around the Flight Director console, and the discussion was quite animated. My interpreter was trying to translate the air-to-ground loops, but said things were happening too fast and the phrases didn't make sense. There were comments about "seal it off, seal it off" and "where is it going?" "We can't see it, we don't know where it bounced."

There was a problem. That was clear enough. Then I heard, "The pressure is still dropping." That could only mean something very bad had happened, and that the *Mir* had been punctured. I concentrated on finding the atmospheric pressure on the displays on the big screens in the front of the control room, and I listened to my interpreter as he tried to sort things out. Eventually, it became clear that the Progress ship had bounced off the *Mir*, hitting the Spectre module (where most of the US science experiments were stored and conducted), and had put a hole in it. The crew had managed to close the hatch but had not saved much of anything inside the Spectre. The pressure in the rest of the *Mir* stabilized at a low but livable value. The next thing I knew, I was being ushered in front of a large array of TV cameras, as it was discovered I was the senior NASA operations guy in the building.

I really didn't know a lot at the time, of course; the Russians were still trying to figure out what had happened themselves. But I knew that the pressure was stabilized, and that Mike was okay. As always with the *Mir*, they had a Soyuz attached. It could be used to evacuate if the *Mir* became uninhabitable. The Soyuz was fine too; they'd checked that right off the bat. The emergency was over. I said a few things for the cameras about having faith in our Russian counterparts, that they understood how to fly their vehicle, and that they would be coming up with a future plan—with our teams included—and we would be following closely. In essence, I really didn't say much except that we would stay involved and that we were happy that the crew was safe. It turned out that there was a tremendous amount of work to do. It required an internal spacewalk to try and find the leak, and all sorts of different procedures were needed to work around wires that had been cut in the haste to close the hatch. It actually was a good lesson in dealing with the aftermath of a space emergency, and it was an outstanding lesson for our teams to be involved in.

Shuttle-Mir was, for most of the MOD, much more about the Shuttle visits than it was about the long-duration astronauts living on *Mir*. For those of us planning and flying those Shuttle missions, the long-duration crewmembers were passengers to be delivered and passengers to be picked up. They were our friends, of course, so they were much more than cargo. Since a large part of their science was about how their bodies reacted to long-duration exposure to microgravity and the space environment, it was important to treat them in a way that preserved that science until the doctors had done what they needed to do to them after getting them on the ground. We even developed a supine seating system so that crewmembers would take all the entry loads through their back (referred to as "eyeballs in") rather than down through their feet ("eyeballs down") with the standard seating system. We made sure that we had a crewmember on the Mid-deck to watch over them during entry, and we had to plan for emergency egress for what we called a deconditioned crewmember. Deconditioning occurs when the human body spends a long time in free fall, and the muscles and bones weaken because they are not being used. The problem was that the Launch and Entry Suit that all Shuttle crewmembers wore during that time frame, along with the parachute and survival gear, weighed close to 90 pounds, which might as well have been a ton for someone who had been floating in a weightless condition for six months. Emergency egress sort of meant being carried/rolled/pushed over to the hatch by the more able-bodied crewmember, having them hook the individual up to the parachute, and then shoving them out the hatch. We just decided that it would be better if we didn't get into a situation where a bailout would be necessary.

Of course, in order to transfer crewmembers, their supplies, and their science hardware to the *Mir*, the first thing we had to do was get docked.

The first rendezvous with the *Mir,* on STS-63, was the product of a great deal of education on both sides. It required a lot of trust building and negotiation. There was a great deal of concern about plume impingement and contamination. Even if we didn't damage the solar arrays physically by making them flutter, there was still the possibility of molecular contamination, which would cause the arrays to degrade. Everyone is familiar with their car getting a little dusty when it's been parked outside for a while. You hardly even notice it until you streak it with your finger, and the dust washes right off. But if you let just a little bit of anything get onto an array, it can degrade the performance by a small but measurable amount. And because spacecraft are always optimized for the most performance and the least weight, the arrays must operate at near-design efficiency over the long haul in order to keep the spacecraft viable.

The fact that the *Mir* was planned to be in orbit for years, and because there was no way to service the arrays, even the tiniest bit of contamination could have long-term effects that could shorten the station's life. It's all about the details, just like anything else in spaceflight. So, many, many people studied the dynamics of thruster plume particles, how they moved, and where they went. The plumes from the Shuttle were then analyzed and modeled, and we shared this data with the Russians to get their agreement and buy-in for our plans. Everything was documented, and the engineering and science were agreed to in the rules and plans. But when we ended up having jet problems on that first Shuttle flight to *Mir,* it all came down to relationships—the trust built up over several years of working together (and probably some good old-fashioned drinking time).

Bill Reeves, a long-time Flight Director, was stationed in Moscow's MCC (known as TsUP) during that flight. He was there because, out of all of the members of our office, he just had the kind of personality that helped him build a social relationship with the key guys on the other side of the pond. Put another way, he knew how to drink with them, and he knew how to relax with them. Bill spent quite a few years assigned to Aircraft Operations during the Skylab days, and he developed a good Officer's Club manner. The Russians liked that. They wanted to see you with your hair down, and Bill knew how to do that. So when the jets failed, and we were looking at the loss of the close approach to *Mir*, he was able to pat his friends on the back and have some heart-to-heart talks that got them to trust our alternate plans. Sometimes you don't get to trust someone just by holding meetings with them in conference rooms.

Decision-making during the Shuttle-Mir Program always had a bit of-behind-the-scenes drama. The Phase 1 Program Office was developed to run the project after the initial meetings between technical groups. Because the Russians preferred to build trust and friendship via individual-to-individual relationships (rather than by simply having meetings between whomever was in a particular position at the time), discussions on major issues usually occurred on phone calls or in small-scale, face-to-face meetings. The NASA way was to hold large, open board meetings—the Russians rarely worked that way. The net result was that there was a bit more mystery to how and where decisions were being made throughout the program. There was nothing evil about it, and for the most part good, sensible decisions were made. But there was a certain uncertainty about which way things might go on various topics. For those used to working in an open operations environment, it could be a bit unsettling. Decisions got made, and no one knew where they came from. It was just sort of the Russian way.

I always liked to say that flying a space station wasn't as dangerous as it might seem to those who flew ships up and back from orbit. After all, the only real quick killers were a loss of cabin pressure or a fire...and *Mir* happened to experience both during the time we flew with them. Less critical, but equally threatening in a programmatic way, was loss of control...and that happened on numerous occasions as well. About the only thing that we didn't experience together during Shuttle-Mir was a toxic atmosphere. Then again, the fire did a good job of putting the crew's breathing systems at risk.

It was always to be remembered that when there was no Shuttle docked to *Mir*, we Americans were mostly observers. Yes, we had a huge interest in how problems were solved. And yes, we had a majority vote in what our crew did or did not do. But solving problems with the *Mir* systems, or the overall direction of their program, was the Russians' right and responsibility. We learned a tremendous amount about how they thought as they worked through the issues, and it was a great learning environment for us as we watched our future partners. They, no doubt, observed us too. Our teams in Russia were usually able to sit in on key troubleshooting sessions because of our interest in the American on board, and we always welcomed the Russians spending time in Houston when we were talking about Shuttle-Mir.

The fire on *Mir* was caused by a failed oxygen generator. It proved to be a huge threat to the future of the joint program (after we determined that it was not an ongoing threat to our astronaut on board). There was a great deal of discussion at very high levels over the amount of risk we were taking with our astronauts' lives on board the *Mir*, mostly because the fire was so dramatic. But as engineers, we sat down, understood the problem—along with the Russian response—and figured a way to justify the risk for the rewards of long-term cooperation with the Russians. Whenever a major failure like this occurred, most of the discussions about determining the cause of the incident and the future of the program went on at a programmatic level. Our operations team meanwhile kept flying on a daily basis. The guiding principle of the folks on the front lines was to continue what they were doing and do it well for as long as they could—until they were turned around. We obviously participated and contributed to the investigations, but it always was remembered that no matter what, if you were in flight, you had to keep flying. It was an old lesson from aviation—no matter what, you first fly the airplane, then you do whatever else you need to do. "Aviate, Navigate, and Communicate—in that order!"

The first docking mission was STS-71, with Robert "Hoot" Gibson in command of the *Atlantis*. The spacecraft was equipped with the Russian APDS—the Androgynous Peripheral Docking System—that allowed any spacecraft so-equipped to dock with any other that had the same system. It featured no probe or drogue—just three docking petals that interlocked with the same type of system on the other vehicle. Capture latches were incorporated into the petals, and the rim included a series of latches that drove from either (or both) sides to bring the sealing rings together for hard dock. The docking systems had to be aligned within about 3 degrees, and within 3 inches of centerline, in order to ensure a good soft dock. Docking velocity depended on the mass of the vehicles being docked—bringing an Orbiter in to dock with the *Mir* was a slow process because a Shuttle had lots of momentum: its high mass required a low velocity. Docking a Soyuz was a more sporty affair because its mass was low—you needed a higher velocity to get the same momentum. In fact, if you were used to docking a Shuttle, with its stately final approach, and you got into a Soyuz simulator to shoot the same approach, the closure speed was positively terrifying. A Soyuz would move at a good solid foot per second, whereas the Shuttle approached at less than a tenth of that figure. It may not sound fast to you, but with a Soyuz there is not a lot of time to make fine corrections to that narrow cone of acceptable angles and offsets in the last three seconds.

STS-71 was a mission that featured a lot of interesting firsts, including the fact that it landed with more crewmembers than it launched with—and no, there were no babies born on board. The mission featured an actual *Mir* crew change. The Shuttle brought up two new cosmonauts, and it brought back the two cosmonauts who had spent their mission on board with astronaut Norm Thagard, who also returned on *Atlantis*. In order to exchange crews, STS-71 had to bring up new seat liners for the Soyuz. Each person who rides up or down in a Soyuz is strapped into a fairly bulky, form-fitted seat insert. This was the first time most of us had been introduced to these liners, but hauling them up and down became a way of life as we changed out crewmembers. Even those scheduled to go up and down on the Shuttle had to bring up a custom-made seat liner, because the Soyuz was their escape craft for their stay on *Mir*. They had to have a safe couch (and a Sokol suit—the Russian launch and entry garb) in case the worst happened and they had to abandon ship.

Fortunately, nothing that dramatic ever happened during a docked Shuttle mission. The flights did not go by without their excitement, however. The most common failure experienced on the *Mir* during these years had to be loss of control caused by computer problems. The *Mir* had an attitude control computer that had three "lanes"—their word for what we would call channels. In reality, it was three computers operating in the same box. Similarly, the Shuttle used four General Purpose Computers operating in sync for ascent and entry. When the *Mir* lost all three lanes, it lost control. And that meant sooner or later (usually sooner) it would lose its ability to point the solar arrays at the sun. This in turn would mean a power reduction and effectively cause a brownout—or, more commonly, a blackout. When the Shuttle was docked, it could control the attitude of the stack. That is exactly what we did most of the time. But first you had to actually get docked, and you needed *Mir* to be in control for that.

Several Shuttle missions launched while *Mir* had known computer lane issues. We worried about additional failures that would leave the station in free drift. Such failures would make docking extremely difficult, if not impossible. Although we developed techniques for doing the maneuver, they wouldn't work if the *Mir* was spinning above a certain rate in any or all axes—and when I say spinning, I'm not talking about some Hollywood-like special effect where you can see it doing pirouettes. Anything more than a degree per second is really pretty racy—although it might not look like it to an outside observer. Fortunately, we never had to fly an actual docking with the *Mir* out of control—but we'd have given it a go if the need had arisen.

Both the *Mir* and the Shuttle used essentially a 14.7 psi nitrogen and oxygen atmosphere, which made it easy to keep the hatches open all the time they were docked together. This open hatch policy was good because the primary goal of any of these missions was cargo transfer. It was our job to bring up as many supplies and science experiments as we could, and it was during this period of time that the job of tracking transfer items became a solid part of the Payload Officer's job description. There was always a huge list of things that had to go from the Shuttle to the *Mir*, and another equally long list of things that had to come back. The Russians, up until the Shuttle visited, were very limited in their return capability—the unmanned Progress supply ships burned up on reentry, so they couldn't bring anything home. The Soyuz could only bring back about as much as a person could carry in their lap, as there was very little return stowage space available in the tiny capsule that made it to the ground. But the Shuttle was roomy, and it was designed for reentry with comparatively huge cargos. Anything brought back from inside the *Mir* had to be stowed in the Orbiter's crew cabin, but this work provided practice and spurred the development of cargo techniques that were used all the way through the ISS construction and servicing missions at the end of the Shuttle program.

One of the big cargo elements on almost every flight were Russian black boxes from the KOURS automated docking system. The KOURS consisted of a number of computers and sensor boxes that were mounted in the Progress or Soyuz, and the KOURS controlled the rendezvous and docking of these vehicles. Russians generally used automated docking, with hand-flying as a backup option if the computers were to fail. The KOURS system was developed in the days of the old Soviet Union, and the factory that designed and built them was in Ukraine—a former Soviet state, but now an independent nation. That meant the KOURS had to be obtained from "outside the country," so to speak. Because relations were sometimes strained between Russia and Ukraine, there were times when KOURS devices were simply not available to the Russian space program. Therefore, the cosmonauts busily removed the black boxes from every Progress vehicle that arrived and stored them for return on the Shuttle— sort of like recycling old soda bottles. (As I recall, we never did see our five cents per box, though.)

The Shuttle, by contrast, was always flown manually for docking. Well, manually with computer assistance, of course—it was all fly-by-wire. Many fans of science fiction movies expect that the pilot of a docking vehicle is flying in what we call "all six axes," not only translating in/out, up/down, and left/right...but also pitching, rolling, and yawing the vehicle. While you can do that if you want, it does keep you pretty busy. And when you are docking something as big as the Shuttle to something that has as many things sticking out as the *Mir*, it is nice not to have to worry about, say, bumping the tall tail of the Shuttle into the tip of a solar array because you pitched a little too far. So we used the computers to maintain attitudes, while the astronauts flew the translations.

Letting the computer (autopilot) fly the attitude (orientation) of the Shuttle was actually quite elegant. It knows how it is pointed because the Inertial Measurement Units (IMUs) have been aligned. We assume that we know where the *Mir* is pointed, because they have a similar system, and we told them how to point. So if you assume a perfect knowledge of both vehicles' attitudes, you can bring them together with perfect orientation. Of course, nothing is perfect, so we developed the technique of slowing to a stop just about 30 feet out and using a special alignment target to see if there was any error in the relative attitudes between the two vehicles. The target was designed so that you could actually read the misalignment in roll and pitch. But you didn't use the joystick to line things back up. Instead, you typed the correction into the pointing display, the Shuttle made the minor correction, and in you went—aligned by eyeball. A TV camera was mounted in the center of the docking port hatch so that you could look straight up the centerline as you approached. We had alternate alignment techniques in the event of a camera failure, but they took a little more interpretation than looking straight up the central axis of where you were moving.

One of the things that the joint teams recognized right off the bat when we started to talk about docking Shuttle to *Mir* was that the docking ports on the existing station were a little too close for comfort—the Shuttle really had to get in close to the central node, the solar arrays from various modules, and the KRYSTAL (one of the experiment modules radiating from the *Mir*'s central node). The solution was a special docking module. It was really not much more than an extension tunnel added to the end of the KRYSTAL that put a little more distance between the Shuttle and the stack. This was designed and built by the Russians, shipped to Florida, and launched on STS-74, the second docking mission. It used an APDS on each end, of course. In order to attach it, the RMS (Remote Manipulator System) was used to lift it out of the payload bay and stack it onto the docking port on the Shuttle. The Shuttle then flew in to dock to *Mir*. When we departed, we left it latched on the *Mir* end and detached from the Shuttle end.

A certain amount of momentum was required to get the APDS capture latches to soft dock the module to the docking port on the Shuttle. The RMS really wasn't designed to generate that kind of force, so it couldn't just slam the docking module onto the Orbiter after it was pulled out of the bay. Once again, physics was our friend. It really doesn't make any difference if the docking system is slammed onto the Orbiter, or vice versa. So the RMS was used to align the docking module and hold it just a couple of feet above the Orbiter. Then the arm was commanded to go limp, and the Shuttle jets were fired to ram the Shuttle into the module. Momentum did the rest of the work: the latches captured, and we had a snout sticking up out of the payload bay. The Shuttle docking with *Mir* went on, and when STS-74 departed, the docking module stayed behind. It served as the docking port for all future Shuttle-Mir missions.

The *Mir* was configured with a central node (essentially a ball). Cylindrical modules stuck out of the ball in the cardinal directions. It looked heavy and stout, but looks can be deceiving with spacecraft. In truth, space structures like *Mir* are light and springy. To give you an idea just how light and springy, we were told by one long-duration astronaut that if they were at the far end of one module, and looking through the central node to the far end of another, and someone began exercising on the treadmill, there was enough flexing from the vibration that the far end of the opposing module actually disappeared from view. Now that's flexible! It was more robust than it might have looked though, at least from a systems standpoint. Astronauts who were on board when it lost attitude control (and, therefore, solar power) reported that it would go completely dark. Then when the sun came back, everything just came right back on—fans, lights, pumps, computers—it just kept on ticking. There were no reports on whether the digital clocks were flashing "12:00" though.

Remember that the Russians had ways of bringing things *to* the *Mir*, but they were not able to bring much of anything *back* until we started flying there with the Shuttle. As a result, the *Mir*'s interior became progressively more cluttered as the years went on. Eventually, it looked a lot like a World War II submarine—cramped, with things hanging all over every wall. It would have been nice for us to help them clean it out, but we had lots of experiments and other cargo to return (like Russian docking computers). Our astronauts even added to the clutter, leaving little bits and pieces of their personal collections behind on every increment. Shannon Lucid, the second astronaut to live aboard the *Mir,* told me that when the station finally reentered the atmosphere and burned up on its way to the Pacific Ocean, she thought of her reading library, the books she had brought along to keep her occupied in her spare time. They'd been left there for future astronauts (and they couldn't come back, because they weren't on the "down manifest"). So they went down with the ship, and she was a little sad about that. A waste of good books, but science down-mass comes first.

After the docking module was installed on STS-74, the Shuttle-Mir Program continued. About every third Shuttle flight went to the *Mir. Atlantis* was the primary vehicle until near the end, when *Endeavour* took a turn. *Discovery* docked the final time on STS-91. There were nine docked missions all told, from June 1995 until June 1998. The final mission carried aloft the last Russian to fly in the program, a surprise passenger, Valeri Ryumin, a former cosmonaut who effectively became the manager of the *Mir* flight program in Russia. It would have been the equivalent of sending the Shuttle Program Manager aloft to see just what the vehicle was like. Ryumin hadn't been to space for a very long time, and he was a big guy—we always figured that he was a retired pilot who had no ambition or need to fly in space again. But there he was, going back up. Ryumin didn't do a lot of training for the final trip to *Mir,* but he did enough to qualify for the flight. He had few assignments during the mission, and he spent much of his time simply observing or keeping to himself, according to some of the astronauts on the flight. It was probably one of the oddest incidents in the entire program, one that shows just how autocratic their program could be.

The Shuttle-Mir Program set the stage for ISS in a way that is hard to underestimate. It is difficult to imagine how we would have gotten started with a brand-new vehicle at the same time we were trying to understand the Russian space culture (and they were trying to understand ours). There were fundamental differences in the way we looked at risk and safety. In one of our very first meetings with the Russian operations team, for instance, we asked them about their landing philosophy. We knew that they brought the Soyuz down in the middle of Kazakhstan, whereas we landed our similar Apollo vehicles in the ocean. For us, one of the advantages to the ocean landing was that the debris associated with reentry would never pose a risk to anyone on land. (The service module, which was larger than the crew capsule, separated after the vehicle was committed to reentry, and there was a possibility that large pieces could make it to the surface.)

In the Russian system, pieces of their service and orbit modules could, if they survived reentry, impact land somewhere in the steppes of Central Asia. We asked them how they managed the risk of this, and if they were worried about hitting anyone. They were puzzled at the question, so we had our interpreter explain it again, asking if the risk of hitting anyone out on the steppes concerned them. It was if a lightbulb had suddenly gone off over their heads as they understood what we were asking—and their answer was, "But...they're just Kazakhs!" That gave us a little bit of a window into their post-Soviet feelings on risk, for sure. While we were shocked at the answer, we later learned that they had a very robust view of what was considered to be acceptable risk. In fact, there were risks that we were satisfied with that appalled them. It wasn't better or worse, it was just—different. And this was true about most of the differences we found.

Our respective views on medicine were a source of humor for both sides. When we traveled to Moscow, we were quite leery of getting sick or injured and then having to be treated by the Russian medical community. We had heard things that made us worried about their grasp of medical science. So when we traveled, we each had a card with a set of phone numbers on it clipped to our badges. It was a way to get ahold of a medical evacuation service that was contracted by NASA to get us out of the country and back to "civilization" in case we had an accident or illness. I know that sounds both grim and rude—but the humor is that the Russians essentially felt the exact same way—but in reverse! They wanted to make sure that if they had a problem, they were treated in Russia, by the Russian medical community. It was simply a matter of perspective, and how (and where) you grew up.

When we visited Russia, we were amazed at the way they drove in Moscow. Lane lines were simply suggestions (most of which were ignored), turn signals were for sissies, and aggressive driving was rewarded, not punished. We'd occasionally take a ride with a Russian colleague and came out amazed and a bit shaken that we'd actually gotten to our destination in one piece. The large ring road (a surface street) around the inner district of Moscow was allegedly six or eight lanes wide—but in practice, the drivers made room for about twelve. Yes, there were traffic police, and, yes, they cited infractions. But there were too few of them and too many drivers, so most got away scot-free. It was exciting, taking a ride with a local in Russia—and exciting in a way that you might not want.

We were able to get even when the Russians visited with us in the United States. On weekends, we liked to show them around south Texas, just as they shared their city and country with us. So one weekend, a group organized a driving trip to San Antonio, about 250 miles west of Houston on Interstate 10. The speed limit was always 70 miles per hour on the highway, but the prevailing speed was more like 85 in those years. Our folks managed to snag a few large passenger vans and loaded them with Russians for the trip west to enjoy the sights, sounds, and food of San Antonio—and a good time was had by all. But the following Monday, we found out that the Russians had been pretty terrified by the driving portion of the trip. Sure, they were a bit crazy when driving around Moscow, they said, but at least they weren't going at such ludicrous speeds! Eighty-five miles per hour was about twice what they could ever get to in their crowded city—or with the ancient Soviet automobiles they still drove.

Like I said—not better, not worse—just different!

Within the Flight Director Office, we had a small but dedicated cadre that flew most all of Shuttle-Mir missions. Bob Castle, Phil Engelauf, Bill Reeves, and I worked most of the missions in some capacity or another. It made sense; there was enough overhead in learning the ins and outs of the Shuttle-Mir Program that it paid to have a dedicated team. We had the Russians over to our houses for parties, and they got to know us while we got to know them. Likewise, the key flight controllers—especially the RIO team—worked most of the missions and let the rest of Mission Operations work the other two-thirds of the Shuttle program during that time. Another cadre was getting ready to fly the ISS, which was only two years from first element (the first piece of the space station) launch when we wrapped up STS-91. A number of our RIOs were selected as Flight Directors—the first group of Flight Directors to prepare to fly the ISS with the Russians. In this way, the experience and relationships gained during the Shuttle-Mir Program really paid off. While there were technical challenges along the way, and it was a constant struggle to keep up with failures on the aging *Mir* so that the program could continue, we did our best to support the Russians and we both learned a great deal from one another. But mostly, we learned how to work together, 24/7.

There was a hiatus in Russian-American flight between the end of Shuttle-Mir and the first launch of an ISS component. The behind-the-scenes work and the relationships continued apace, however, and there was never a time when there weren't a significant number of Americans in Moscow, and a similar number of Russians in Houston. This is why the Shuttle-Mir Program name was changed to Phase 1 early on—it was simply the beginning of the ISS, only with different hardware and software. The people were the constant, and the seeds of an international space culture had been sown. They would come to fruition in due time, and it's a culture that continues to grow and mature to this day.

Chapter 8

SRTM and the Mast

LeRoy looked worried. Not that I blamed him; there were a lot of unknowns in what he was doing. I turned up the sound on the TV in my kitchen—it was always on and tuned to the NASA Select Channel whenever we had a mission in orbit—and listened for a minute. Technically, I was sleeping. But as the Lead Flight Director for a mission, you don't get much rest in between shifts, especially when flying a one-off mission like STS-99—the first flight of the Space Radar Topography Mapper (SRTM). The SRTM payload consisted of two large antennas—one that took up the majority of the Shuttle's payload bay, and one that sat on the end of a 200-foot-long mast that extended out of the bay to the Orbiter's left. The mast had to be deployed after the Shuttle got on orbit and the payload bay doors were opened. It needed to be stowed again to get the doors closed so that the Shuttle and its crew could come home. As I listened to the commentary from the Public Affairs Officer, it didn't take long for me to share LeRoy's concern. It was entry day—and the mast wasn't stowing. I realized that I was going to need a tie because there was no way I wasn't going to head into the Control Center. This was my mission, and if we had to make the decision to jettison the mast, I needed to be there. After all, I'd been working this flight for over a year.

The first Shuttle flight of the twenty-first century was a simple-sounding mission that will probably have more of a lasting effect on a greater number of people around the world than almost any other mission flown. It wasn't glitzy or glamorous, didn't have any space walks, didn't deploy any satellites, and didn't launch a planetary probe. It didn't create or service a space station. But what it did do was create a topographical map of the entire Earth—a map that will be used as the baseline data set for all geographical work for the next century. It will help plan transportation and communications systems. It can show people the best places to live and the places to avoid. And it serves as a baseline for educating anyone with a connection to the internet on where everything is on our planet. Human beings have been trying to accurately map the entire planet for as long as human beings have been around, and before we launched STS-99, there were still significant spots labeled "terra incognita."

Our friends at the Jet Propulsion Laboratory (JPL) in Pasadena, California, have always been known for cutting-edge science and the hardware and software to make it happen. They had flown two previous missions with space radars mounted in the Shuttle's payload bay. These were used to learn what they needed to know to fly the big topographic mapper. I was not the first Lead Flight Director assigned to the mission. I was asked to take it over from my classmate, Bryan Austin, sometime about a year and a half before the flight. I had come off Shuttle-Mir wanting a break from working with the Russians. I had been doing it for the better part of the 1990s, and I knew we'd be in it again for the long haul once the ISS started flying. So it was a pleasant surprise to be offered SRTM. It came with a few key players already assigned—for instance, the Payload Officer and FAO. Much of the conceptual work had been done, and the hardware/software package of the payload were well in the works. That meant my first task was getting myself up to speed so that I could keep up with the rest of my troops! There's an old leadership joke: "Which way did they go? I am their leader and must catch up!"

As Lead Flight Director for a Shuttle mission, you are never truly off duty when the bird is in orbit. Oh sure, you work your shift and go home. But when you get there, on comes the TV with NASA Select burned into the tuner. The planning and training for a mission takes so long and is so involved that you simply can't get it out of your head when you leave the

Control Center. The fact is, the Lead is intimately acquainted with the entire flight. The Orbit 2 and Orbit 3 guys—not so much. Sure, they got the briefings, did a sim or two, and watched a couple more. They read the manuals and made sure they understood their role—but the Lead was the one everyone turned to when something went wrong. It was his (or her) job to make sure that all the contingencies had been thought out, that the procedures and rules had been developed and tested. The other guys could execute, but the Lead was the one with the corporate knowledge of why things were going to be done.

It was traditional for the Lead to take the Orbit 1 shift, which on a normal flight amounted to the crew's morning. Major activities were generally planned to start first thing in the crew's day. This allowed for plenty of time to fix problems if they appeared. The Orbit 2 Flight Director took the crew's afternoon, and the Orbit 3 Flight Director was on while the crew slept. This was the planning shift. If everything was going the way it was supposed to go, the planning shift could be quite peaceful. On the other hand, it could be insanely busy if the entire plan for the next day needed to be rebuilt due to ongoing problems. There were times, of course, when the vagaries of orbital mechanics or night and day cycles conspired to put major mission activities on the Orbit 2 shift—in which case the Lead might choose to take that time slot instead. The Ascent/Entry Flight Director popped in for those dynamic phases, of course, but was often blissfully unaware of the details of the payload—other than what it weighed and if it was hazardous to atmospheric flight.

Things got trickier on a mission like SRTM. This was a science mission that had to run twenty-four hours a day from start to finish. The crew was split into two teams (red and blue), rotating from sleeping in the Mid-deck to operating the spacecraft and payload recorders on the Flight Deck on twelve-hour shifts. There were no unimportant hours on this mission, so as the Lead I had to make sure that my counterparts on the Flight Director console were well enough trained to keep the ball rolling no matter if I was there or not, or if I was sleeping.

Planning for the mission was well underway when I was assigned the leadership role, and the last thing I wanted to do was stir the pot in a way that was counterproductive. It was getting close to the time for when the first payload simulations would take place at JPL, and I stopped by there a couple of times to meet the team and let them educate me on the payload itself. There were some remarkably smart people working on the team, and I am not just talking about the engineers and operations gurus. The folks who had figured out the science behind the radar were well above my creativity level, and I was just in awe of what they had created and just how detailed their analysis and design turned out to be. They appeared to be accounting for every potential error—misalignment of the external antenna, for one. They had created a laser ranging and direction system that determined exactly where it was relative to the inboard antenna at all times. They didn't have to drive the antenna to some perfect position; they could measure where it was every second and correct the data mathematically to make it accurate.

The mast itself was a leftover from the Space Station Freedom program. It was originally a solar array support mast but was no longer needed for ISS. It was an amazing self-erecting Tinkertoy set (in fact, there were models in some people's offices made of actual Tinkertoys to help them understand the structure) that was 200 feet long when deployed but fit in a canister about 10 feet long when stowed in the payload bay. Deployed using dual motors, the segments snapped into place as it cranked out of the canister, and it likewise snapped back into a compact stack when stowed.

Our crew had been assigned for a long time, and they had prepared carefully for their respective roles. The commander and pilot (Kevin Kregel and Dom Gorie) had trained carefully for some special maneuvers that would be needed to keep the Orbiter at exactly the right altitude each day. The Mission Specialist team—Janice Voss, Janet Kavandi, Gerhard Thiele, and Mamoru Mohri—were experts on the radar systems. They were ready to handle any malfunctions that might arise, but they joked that their main purpose was to continue feeding tapes into the high-rate data recorders located on the aft flight deck. All the radar data was recorded on board, and only some was downlinked to the ground during the mission, just so that the Payload Operations Control Center (POCC) could check to make sure that the instrument was working properly and the data we were collecting was good.

The purpose of the STS-99–SRTM mission was to map the entire Earth in 3-D to a resolution of 9 meters and an accuracy of 1 meter. That basically meant that if you covered Earth's entire surface with a 9-meter grid, we'd know (when we were done) the altitude of each of those grid points with an accuracy of 1 meter. Of course, we'd only get Earth's surface between the north and south latitudes of 60 degrees—so forget about Antarctica—but the parts of the world where most people live would finally be known with great accuracy. I often stopped to think about what that really meant. I had been a fan of exploration all my life, and I have personally spent much of my life exploring rugged wilderness areas (and am a Fellow in the Explorers Club). One of the fundamentals of exploration was developing a map of the areas where you visit. The human race had been trying to map Earth's surface since humans have had the intellectual capacity to understand geographical relationships, and now we were finally going to make the dream of a world map come true. No more terra incognita, no more "here be dragons" for the odd, hard to reach locations of the planet's surface. We would *know* the surface of our planet.

Not only would we be able to use the big radar to generate this map, we were going to do it in just ten days. The smart folks at JPL who developed the mapper had calculated that if we flew at a specific altitude and inclination, and mapped continuously, we could cover every spot on Earth's surface with the radar beam three times. Getting repetitive passes over the same piece of ground improved the accuracy, especially since each view would be from a slightly different aspect. One hundred fifty-nine orbits would create a complete map, provided we didn't miss any passes and did things like change the tapes and make trim burns while we were over the ocean. The trim burns were small maneuvers to account for increased atmospheric drag at the low altitude we were flying, intended to keep the orbit at exactly the right distance from the earth for an accurate map.

The mission was planned this way from the start, hence the need for twenty-four-hour crew operations and constant vigilance of the orbital parameters to make sure we kept that altitude perfect. The SRTM used a lot of power, and unmanned satellites that relied on solar power could not provide the sort of energy levels required by the big antenna. The Space Shuttle was the best choice as a launch platform, and the fuel cells of the Shuttle could give SRTM many kilowatts, provided that the cryogenic oxygen and hydrogen held out. These were the limiting consumables for the mission, and we didn't have a lot of extra with the tankage on board. During preflight planning, propellant was not considered the limiting consumable. It turned out to be, though. The bottom line was that the mission was planned with full cryo tanks, and the reality was we would have just enough capability to support the full map. There was little margin, other than the deploy and stow operations.

Generic flight rules for a Shuttle mission always provided margin— it was the best way to give yourself flexibility in contingency cases. In general, we always wanted to be able to come down out of the sky in an emergency, and to do so quickly. Most systems were designed to support this emergency landing capability. The payload bay presented unique problems in this respect. The big doors needed to be closed for entry, and because many motors and latches needed to work to get them closed, we liked to provide enough time in the plan to deal with failures. There were always a few things that further complicated closure of the doors. For example, a device that needed to be stowed might project beyond the payload bay "moldline." This problem would need to be solved before you could even get to the point of closing the doors. The Ku band antenna, for instance, was such a device. A more obvious example is the Remote Manipulator System (RMS)—the robot arm. It was always hanging out, and it needed to be dealt with. Both could be stowed with their nominal systems in a few minutes. If that didn't work, you could plan for an EVA (Extravehicular Activity) to go outside and stow them manually.

And if you didn't have time for that, both could be jettisoned with the use of guillotines and explosive bolts.

Because hardware costs a lot, and EVA takes time, the generic rules called for the stowage of all such items the day before the nominal entry day. In other words, you never wanted to find out that you couldn't stow the antenna or arm on the day of entry because in order to try the EVA to save the hardware, you'd need to abort the deorbit attempt and eat up one of the contingency days that you'd rather save for weather problems. So you always planned to stow these items the night before entry. As a result, if there was a problem, then an EVA could be inserted more easily into the timeline, and the planned landing could still take place. This rule was assumed to be valid for any experiment or satellite operation that extended beyond the payload bay sill. However, it was not an absolute rule, and in the case of the SRTM we had planned the mission such that we would leave the mast deployed until the map was complete. This pushed us right up against the deorbit time frame.

Why did we take such a "reckless" position? Well it was simple: the goal of the mission was to bring back the map not the mast. The mast was simply a tool to get the map done. Assuming that the map data was collected, then the mast was expendable. It was no longer required, no longer needed—except as a museum piece. It was never planned to fly a second time and, in fact, was certified for only a single mission. While the manned Shuttle program was built on the fundamentals of reusability, the Jet Propulsion Laboratory was used to a different paradigm, which was that hardware was expendable.

The Space Shuttle program was big, and it encompassed many different people working on a lot of missions. Due to the large scale and limited resources, it was common for the operations team to plan flights, while the program manager and his immediate senior staff gave them little attention until they were "next up." The program was always represented during mission planning by the Mission Manager who worked directly for the Shuttle program and who, indeed, was responsible for bringing all the hardware, software, and plans together from all the various centers and organizations that participated in a Shuttle mission. In fact, it could be argued that the Mission Manager had greater overall insight than the Lead Flight Director. The Lead Flight Director was responsible for the flight operations planning, training, and flying, but it was the Mission Manager who had to ride herd on all the various elements of hardware and software that had to be processed at the Cape, at Johnson Space Center, and at the payload centers that contributed to the mission. The Mission Manager's work was largely over when the vehicle left the pad, whereas the Fight Director's job was kicking into high gear. It's worth noting that the Mission Manager himself (or herself) might very well be working several future missions at one time and, as such, had to triage their own time and decision-making.

In those days, the head of the Shuttle program might first give serious attention to the details of a given mission during the program review of the Flight Rules Annex (the rules specific to a given mission). This review might be conducted in the last month before flight. The program must sign the flight rules, which effectively become a "contract" between the Shuttle program and the Mission Operations team. These flight rules define how the team will operate to execute the program's mission. The rules also try to anticipate potential contingencies and provide a preplanned path through the woods that the operations team will use absent further discussion with the program.

When the SRTM rules made it to the desk of Shuttle Program Manager Ron Dittemore, it was probably the first time he had spent a lot of time thinking about them, and he didn't really like what he saw. The SRTM mast stow time was at issue. In his mind, by waiting until the last minute to stow it we were risking the chance of either jettisoning the payload or throwing away an extension day if there was a problem. It was an untried deploy and stow system and, therefore, he felt that we should be more conservative. The first I heard of this redirection was at a meeting with Bill Gerstenmaier, then Ron's deputy for operations. "Gerst" was (and is) a pretty smart cookie. When we sat down and I explained the rationale behind the plan to leave the mast out, he asked some good questions and, in the end, agreed with me. The bottom line was that JPL didn't want the mast back—they wanted the data. Assuming that they got the data, the mast was simply something that had to be disposed of postflight. As I put it in several meetings, JPL *never* got their hardware back—they were used to sending things into space and letting them go.

The problem was, Ron had fairly well made up his mind that he was right, and Gerst was stuck in between. The solution that he offered was actually fairly smart. We all knew that every mission "built margin" once it left the ground. The nature of consumables management is that you always slightly underestimate need and overestimate what it will take to complete the operations. Because of this, when you actually get into real time, you always grow margin because the actual usage is less than planned. In the end, we came to an agreement with the program that allowed us to keep the plan to leave the mast out for the entire mapping mission *if* we could build enough consumables margin to give us the additional day at the end such that we could maintain the two entry contingency days that we always saved.

Of course, the best-laid plans always have a way of coming back to haunt you. One of the interesting physical aspects of hanging an 8,000 pound object on a 200-foot-long mast that sticks out the side of the Shuttle was that the vehicle would not naturally want to stay in the mapping attitude. All objects in orbit that are not purely spherical in shape are subject to Gravity Gradient (GG) torques. Think of it a little like a pendulum that wants to seek a natural low-energy state—it wants to hang straight down. "Lumpy" vehicles have attitudes in which they want to settle if there are no jet firings to hold them in other attitudes. In this case, the outboard antenna was going to drag the Orbiter to a left-wing-down attitude in short order. Equally unfortunate is that the attitude control thrusters on the Orbiter were all on the fuselage, meaning that the Shuttle had very little torque capability in roll. In other words, it was going to take a lot of Orbiter gas to keep the vehicle in a mapping attitude. (Basically, the mast had to be horizontal, and GG torques would make it tend toward vertical.) For this reason, propellant would be the limiting consumable if you couldn't design a solution.

The smart folks at JPL knew this, of course. They had come up with a simple, elegant scheme to counteract the GG torques—a cold gas thruster. Basically, they put a tank of pressurized nitrogen in the payload bay and connected it to a tiny thruster out on the end of the mast through a valve, pressure regulator, and a long hose. The thruster was sized so that it was essentially a small leak, pointed in the correct direction to exactly counteract the GG torques. It was simple, straightforward, and almost foolproof. It was that "almost" that got us. On Flight Day 1, with the mast successfully deployed, we saw that the thruster was doing a good job of counteracting the torques. Then we saw a change: The thruster was giving *more* force than needed to balance the mast, and after a couple of hours we started firing jets to keep the mast from heading vertical. The thruster had failed. What had actually happened is that the simple regulator had frozen in an open position, dumping the entire mission's worth of N_2 in a couple of hours, and leaving us to manage attitude with nothing but the Orbiter's vernier thrusters. Suddenly, our problem wasn't cryogenics, it was propulsion.

In contrast to the very simple cold gas thruster solution to the Gravity Gradient torques, the dynamics of the mast itself presented a very complex problem—one that we had worked on for better than a year to solve. The mast was a self-erecting four-longeron structure with diagonal cross braces. The 200 feet of mast folded and unfolded itself from the canister that fit across the payload bay. Interestingly, it was actually the same design as the masts that hold the ISS solar arrays. While it was remarkably stiff, there was no way that it was stiff enough to hold the outboard mast in perfect alignment when outside torques were present; everything has a certain amount of flex. The good news was that it tended to return to the exact same location when vibrations damped out—after, of course, it had been deployed and the low-energy position had been attained. There was no guarantee that every deployment would end up in the same place.

In order to understand the radar returns and translate them into interferometer data to determine terrain elevations, the relative positions and attitudes of the payload bay antenna and the mast-mounted antenna had to be known precisely. To do this, a laser system, mounted in the payload bay, was pointed at mirrors and retroflectors mounted on the outboard antenna. This laser system tracked the position of the outboard antenna constantly and the data was recorded along with the radar data so that post-processing could precisely locate the antenna at each moment in time and make the appropriate corrections. Raw data from the SRTM was of no use, but post-processed data was highly accurate and valuable.

In any case, it was important that the mast not experience a lot of deflection while it was deployed. Something like a reboost trim burn (using Orbiter jets to kick the orbit back up each day after atmospheric drag had lowered the orbit), or even using just the aft-pointing maneuvering jets instead of the larger Orbital Maneuvering System engines, were enough to send the mast rocking back and forth like a fly rod that had been shaken. It was that fly rod analogy that gave us the key to the problem. Because we were flying at an unusually low orbital altitude, the drag was significant. It meant we needed a trim burn of a couple feet per second each day. That worked out to about ten maneuvers throughout the flight, each one needing to be done in a way that we didn't disturb the mast—deploying and stowing it would take hours.

If you watch a fly fisherman in action, it appears that at times they can freeze the motion of the rod while it is bent by moving their arm at exactly the right rate after achieving the bend. Some of our smart people decided that if they could time it right, a small pulse of the jets at the beginning of the burn would deflect the arm backward, then—just as the stored energy would start to make the mast move back to center (where it would pass through and deflect the other way, then come back, then forward...)— starting the main burn would hold the mast in its deflected position while the main impulse was imparted to achieve the necessary velocity change to raise the orbit. Then, at the end of the burn, another pulse of the jets could be used to stop the mast as it deflected forward. So instead of it oscillating back and forth, disturbing the alignment of all the various pieces, it would bend back, stay there, then bend forward into its neutral position— with no oscillation. Fly-casting in space had been born!

The details of the fly-casting maneuver were studied using a variety of engineering simulators and various math models. Once the parameters had been worked out, our people moved on to developing flight procedures for the crew, and began testing them in the various training simulators. This work had been well underway when I was assigned to the SRTM mission, the primary developers in the Mission Operations world being the Lead FDO (Flight Dynamics Officer), Chris Edelen, and the Lead GNC (Guidance, Navigation, and Control), Mike Sarafin. Their work was instrumental in making the mission happen. Both ended up being selected as Flight Directors in later years.

A day in the life of SRTM generally included one of these fly-casting maneuvers to trim the orbit back up. These burns had to be scheduled over long water passes so as not to miss areas of dirt that we wanted to map. Dirt passes were intense with mapping by the Payload Operation Control Center, and long water passes were intense for the flight control teams as we took care of burns and other housekeeping chores that could interfere with the mapping. It was a nonstop mission from start to finish, with continuous replanning efforts as the red and blue astronaut teams alternated sleeping and being on duty. Hanging over the entire mission (especially after the cold gas thruster failed) was the need to conserve prop and cryo to try and make the extra day so that we could leave the mast out for the total mapping mission.

Of course, with the mast hanging out and the desire not to disturb it, Orbiter attitude control had to be done with the small vernier thrusters—two on the nose, and two on each side in the tail. The vernier system was intended for fine control of attitude, while the larger primary thrusters (with about ten times the thrust) were designed for major maneuvers. Verniers may have well been an afterthought in the Orbiter design, because they always seemed to be an underdesigned system. Now all the Orbiter thrusters used hypergolic propellants—a fuel and an oxidizer that spontaneously combust when they come in contact. This is a wonderfully simple system for a thruster, because all you need are two solenoid-operated injector valves and you get thrust. The only real problem with this simple design is that if you get a little contamination in the valve seat, you get a leak—hopefully of only fuel, or only oxidizer, so you don't get combustions. When you get a leak, you get evaporative cooling and the temps drop—freezing the thruster. Of course, there are heaters to keep the valves and thrusters warm, but heaters can fail. And in the case of a leak, they might not be able to keep up with the cooling effect of the leak.

Now the vernier system, being a little bit of an afterthought, had temperature transducers that were limit-monitored in software running in the Shuttle GPCs (General Purpose Computers). The limit values were, in fact, hard coded into the software. This meant we couldn't adjust or play with them in the case of a bad transducer. If you had a leak indication (cold temp), you couldn't use the thrusters. And since there was no redundancy in the vernier system, if one thruster leaked—or if one transducer shifted—you were pretty much done with verniers. Now it was much more common to have a transducer problem, or for a shadow to fall on the thruster (which got it only slightly colder than the limits, though it was still fully functional), than for an actual leak to occur. We dearly wished that we could fiddle with those limits—but the only way to do it was with an actual software patch. Generally speaking, a patch to the Shuttle's primary software in-flight was, at a minimum, a twenty-four-hour proposition. You couldn't afford for someone to make a mistake in the process. But if we had a cold thruster that tripped the limits for STS-99, a wait that long would cause a loss of a significant portion of the mapping mission.

The answer, of course, was to have the appropriate patches to take out those limits preapproved and canned, ready to implement. We had well-thought-out criteria documented in the flight rules that everyone agreed to in advance. They defined the conditions under which the patches could be implemented, and everyone signed up to them. That meant that if we had a vernier temperature transducer issue, we could put the fix into work immediately and solve the problem quickly without consulting anyone for permission. Now I have always told flight controllers that the more failures you prepare for, the fewer will actually happen, because what you prepare for will never happen. But in this case—that wasn't true.

There was one orbit each day that covered the longest stretch of water you could get in our inclination. You came off the Kamchatka Peninsula into the North Pacific, traversed the Pacific to the southeast, ducked almost under South America, and didn't make landfall until you hit the equator on the western coast of Africa. It was something like a fifty-minute water pass. And on Flight Day 7, as we were coming off Kamchatka, I heard, "Flight, PROP—looks like we've had a transducer fail on F5R." I looked up to see the message that denoted the failure of verniers, and sure enough, we had dropped into free drift—the Shuttle attitude was not being controlled. The mast was safe, we just weren't going to be holding attitude. A quick discussion confirmed that we were in one of our preplanned patch cases, and I gave the DPS Officer a Go to put the patch to work. This, of course, required coordination between all the operators on the left side of the room—DPS, INCO, PROP, and GNC. There was little I could do but sit back and watch the clock, the world map, and the team.

When you know you have built a good team, the best thing you can do is trust them and stay out of the way. I made sure that they knew how much time we had, and then I watched as they went to work. Patching the software in the Shuttle GPCs was never something you did lightly. It amazes most technically knowledgeable people to find that the individual computers had only 256K of RAM—and that was double what they had on the first flights! The small internal memory meant that the software had to be incredibly compact and tightly written—there were no spare bits. When we'd do a patch like this one, we needed to change a single word. And we had to know exactly where that word was stored because we were going to address the memory directly. The patch had been simulated on the ground, of course—tested and verified that it would do what we wanted. But if we were off by one bit on the memory location, we could be writing to something else entirely and bring the entire machine to a halt.

Getting the patch right was all about checking, rechecking, and triple checking—and in this case, we had to do it with a tight timeline. But I was happy with the team, and they knew their job. They used all the time they needed, took no chances, and triple checked their work. Not only did we get the vehicle safed, the patch loaded, and the thrusters all reconfigured, we maneuvered back into attitude before hitting the dirt and didn't miss a pass. It was incredible teamwork. All I had to do was sit back and watch the folks that I trusted do their jobs—perfectly.

Back when we did our first simulation for the SRTM mission along with the Payload Operations Control Center (POCC), we had a few problems. Well, actually, we had quite a few problems. Simulator problems, simulated problems—and problems caused by our own making. It seemed like every time we came "feet dry" to start a data pass over land, there was something wrong with the SRTM, or with the POCC's commands, or the POCC's plan—and as a result JPL was missing valuable observation time. It really can be overwhelming, to be honest, to find yourself in the middle of a complex Shuttle mission with brand-new hardware and software. Time is required to shake this stuff out. In fact, from the time we started playing with a new simulator or software load, it usually took hundreds (if not thousands) of hours to get it running properly and behaving the way the vehicle would. We flew the Shuttle over and over again, so our experience base built up mission after mission, and our simulator, procedures, and training kept getting better and better. Unfortunately, the POCC and SRTM team did not have this same background, and it showed.

When the sim was over, and the debriefings complete, I had a chat with the head of operations for the SRTM team at the POCC and expressed my concern that his folks kept missing their cues and not being ready. "Well," he said, "you have to realize that we have a lot going on, what with getting the hardware and software ready for the mission. I can't expect my folks to be perfect."

Now I understood where he was coming from, and I understood his problems. I also understood the culture of JPL in the sense that they were used to taking time to do things with slow-acting spacecraft on the surface of (or in orbit around) some distant planet. But flying a Shuttle payload was different and I couldn't afford to waste the limited sim time we'd have for the mission—or to come back with an incomplete map because we weren't ready. So, I simply said, "The fact that you don't expect them to be perfect is exactly why they won't be! My folks know that I expect them to make no mistakes, so they don't. Your team will live up to YOUR expectations. So set them high!"

He must have taken my comments to heart; either that or he was so angry with me that he did whatever he needed to do to get his team whipped up. Regardless, the teams improved as the simulations went on, and by the time we were ready to fly we were confident in our ability to get the entire map. The first flight day had a jam-packed timeline. We needed to get the mast deployed and checked out so that we could get the mapping started as quickly as possible. The deploy and checkout went fine, with only a few surprises. In any big experiment like this, unexpected things happen when devices are first turned on. The SRTM was no different, but the POCC was quick to recognize the difference between "I didn't know it would work like that" and "It's broken." Fortunately, nothing (except that dang cold gas thruster) was broken. By the time I came in for my second day of the mission, the mapping was underway.

The teams in MCC (Mission Control Center) were mostly working to maintain the perfect orbit and the perfect attitude—everything we did was directed toward that specific goal. In addition, we were working hard to "make prop" (make propellant) and conserve cryo. Of course, we weren't actually making any new propellant. Instead, we were constantly trying to find ways to save it so that we built a surplus as the mission progressed. The goal was to make more margin over what we had expected to use. The same thing was true of cryo—the stuff that made electricity. Orbiter systems were managed during the mission to draw as little power as they could so that more was available for the radar. Again, the goal was to get the extra day in the bank so that we could get the entire map.

The Trench and GNC folks were keeping track of the altitude and making sure that we stayed in the specific desired orbit at all times. To do this, they carefully planned the fly-casting burns for each day. They made sure that each one would get just the right amount of reboost to make up for the orbital decay we were seeing at the low altitude we were flying. Kevin and Dom had perfected their fly-casting procedures and techniques. As long as we got them good burn targets, they made sure that the orbit was perfectly maintained and the mast stayed in alignment.

Meanwhile, the POCC was busy—and the walls of MCC began to show it! I don't know where they found the budget for large-scale flat-bed color printers, but they must have bought a lot of them! By the third or fourth day of the mission, we were beginning to see 6-by-4-foot map printouts, in full color, popping up in the hallways. I don't know if it had anything at all to do with the science, but it sure did inspire the troops to do a good job! There were false-color images, full color images, black-and-white images, and images done with an isometric view—essentially 3-D. The POCC even distributed 3-D glasses to all the MCC teams so they could see the amazing 3-D pictures hanging on the walls in all their glory.

In smaller scale, each shift prepared a map of the swaths that had been covered to date. They were cumulative, and we kept them in a stack on the Flight Director console. A quick glance at the latest one showed what had been covered up until that point, and what had still to be done. Three passes over every bit of dirt were needed to get an accurate map, and there were only a couple of tiny spots under our ground track that we had missed. They were still available for future orbits though, so spirits remained high that we'd get everything we needed. (Eighteen years later, as I write this book, I still have that stack of maps here in my office—a reminder of the remarkable team, and the remarkable mission we flew.)

At each daily Mission Management team meeting, we showed our progress toward building sufficient propellant and cryo to add the extra day to the mission. About two-thirds of the way through the mission, it was apparent that we'd be successful. This allowed for the full mapping mission, with an extra day in case the mast didn't stow. We weren't terribly worried about that, of course, because the deploy had gone all right.

Of course, the mast had been hanging out in the hot and cold vacuum of space for ten days, so we shouldn't have felt too comfortable. Which brings us back to mast stow day . . .

LeRoy Cain was one of our newer Flight Directors at the time (and went on to be one of our experienced Ascent/Entry types later in the program). He was our Orbit 3 Flight Director for the mission, and the mast stow procedure fell on his shift. Technically speaking, I was done with the mission by this point; my shift would be taken over by the Ascent/ Entry Flight Director, Wayne Hale, for the day before entry. I could sit back and relax in the glow of a job well done and all those maps spread around the building. Of course, that's metaphorical; I never relaxed until everyone was safely on the ground. We also had all that data stored in tapes in the Orbiter—only a little bit of the entire mission data had been downlinked. It's not over until it's over.

When LeRoy's team went to stow the mast, it retracted normally until it got to about the last 6 inches of travel. Motors pulled the mast in, and the mechanism collapsed the segments as they reached the mouth of the canister. When it was all the way in, latches held the lid closed, sort of like a jack-in-the-box toy. For some reason, upper management cringed when I described it that way, probably because we'd all seen jack-in-the-boxes spring open when we were children—not a pretty image if our payload bay doors were closed! Regardless, the procedure was to drive the mast all the way to the stow position, stop the motors, then turn on the latch motors to drive those closed.

The problems started on the first stowage attempt: the mast sucked almost all the way in—and then stalled. The lid wasn't all the way closed, so the latches couldn't grab it. How many of us have ever tried to get all our things back into the suitcase at the end of a long vacation? It can be a struggle. With cables and joints cold-soaked by a week and a half exposed to the space environment, in retrospect it was no surprise that they might be a little stiff. When I saw the stow process on the TV in my kitchen, it had reached this point, and there was a lot of head scratching going on. By the time I had reached the Control Center, pulled out my headset, and plugged in, they had already developed a plan. Basically, they were going to extend the mast a few feet, then engage the stow motors at full speed, essentially ramming the mast into the canister. Then they'd hit the switches for the latches to grab on before it could rebound. Imagine jumping on the bulging suitcase and then having someone ready to pull the zipper (or hit the latches) as soon as they saw it had closed. Same idea.

The good news was that it worked, and I didn't need to worry about it coming home without the mast and outboard antenna. Of course, as we had pointed out to the Shuttle program in advance, the value of the mission was in those tapes stored away in the Mid-deck lockers, not in the mast and outboard antenna. The rest of the mission went fine, and Kregel brought the Orbiter in for a good landing at Kennedy Space Center to end a significant milestone in human exploration of the planet. It has been said that it is hard to explain your explorations when you can't tell anyone exactly where you had been. For this reason, maps are vitally important artifacts of an expedition. And the very-high-resolution topographical maps from this mission are valuable for helping us understand the planet on which we live.

Topography drives the best location for cell phone towers, for instance. Picking the right high point can give you the best coverage, allowing rural areas to be served by fewer towers. I know of cave explorers who use topographical data for clues to locate massive sinkholes that mark possible underground passages. And since the radar penetrated much of the foliage of the land it surveyed, you could literally look through the trees for features that had been hidden from view—intentionally or unintentionally.

Time marches on, and missions come and go. I had quickly gone on to preparing for another mission when SRTM debriefs ended. There were congratulatory notes and meetings, and everyone appeared happy. The mapping organizations had their data and said that it would take several years to calibrate and reduce it all (post-processing, it is called) so that it would be available to anyone who needed it. I began training not only to fly Shuttles to build the space station but to become a Flight Director for the ISS as well. I lost track of my JPL contacts until we lost the *Columbia* in February of 2003. I was running the air search operations there, and we were using all sorts of sensor packages on aircraft to look for vehicle debris in the forests of East Texas.

One airplane had a radar sensor that looked down through the trees and brush, but it kept getting so many returns from old cars, rusted out water tanks, and even the nails in fences that the data wasn't helping us much. All it told us was that metal-age humans had been living in the area for a long time. What I needed was a way to subtract out all of the metal that had been there before the pieces of the *Columbia* had rained down on the area—and I thought of SRTM. Perhaps their raw data showed much of the metal that was there in 2001, and we could use that as a baseline, looking for significant changes to indicate where pieces of the Orbiter might be. Everyone in our search headquarters agreed that it was an interesting idea. But, unfortunately, when I made contact with the radar experts from the SRTM mission at JPL, they said that the wavelengths were all wrong, and that they simply didn't have data that would be useful. They knew that because they had had the same idea a couple of weeks before I did and had looked into it with hope—but it wasn't to be.

The mast? I ran into that thing by surprise. I was visiting the Smithsonian's Air and Space Museum annex in the Udvar-Hazey Center near Dulles International Airport in Washington, DC, a few years later. While approaching the Shuttle *Enterprise,* which was then on display where *Discovery* now resides, I happened to look up. And there, hanging from the rafters, was the mast and outboard antenna for museum visitors to see and wonder about. It was just a piece of old space hardware, of course, one that LeRoy and his team had saved from jettisoning that day in low-Earth orbit. I had always argued that the only reason that JPL would want the mast back was so that it could end up in a museum—and it turned out that I was right. I just didn't know what museum it might be.

Chapter 9

Life in Mission Control

The Control Center is more than just a place where people go to work from nine to five—it has been a home away from home for thousands of flight controllers over the years. Starting with the Gemini and Apollo missions, and continuing into the Shuttle (and ISS) era, Building 30, as it is frequently referred to by the cognoscenti, has grown and changed considerably. The Shuttle era started in the old, original Mission Operations Wing (MOW) that was built in the early 1960s, and then transitioned over to the new wing of the MOW in the mid-1990s. Shuttle operations stayed there until the end. ISS operations have moved back and forth between the old and new wings as flight control rooms were refurbished and upgraded.

The various back rooms that make up the bulk of the square footage and house the majority of each flight control team, have always been spread around the building. Through the miracles of electronic communication, it made no difference if your back room was on the same floor as the front room they were working with—or even in the same wing. Back rooms were variously referred to as Staff Support Rooms (SSRs) or Multipurpose Support Rooms (MPSRs), depending mostly on when you arrived to work in MCC (Mission Control Center). The two terms were created in different eras but referred to the same thing as far as most flight controllers were concerned. On any given day, flight controllers could be supporting multiple flights and simulations, or simply be supporting development work—and all of that could be going on at the same time.

Mission Control was originally built in the early 1960s. It was a mar-
vel of its time. A three-story, windowless building with an attached office
wing, it housed one of the most advanced computer complexes in the
country. Five large mainframe computers chugged away on the first
floor. Many subsidiary computing and communication boxes and racks
filled out the floor, creating the Real Time Computing Complex (RTCC),
the electronic heart of the Control Center. Surrounding the central space
of the floor was a corridor that ran all the way around the roughly square
building. On the outside of the corridor were yet more rooms filled with
electronics and support facilities. The Communications Center was a
long room filled with racks and racks of panels with patch cords connecting thousands of sockets that routed voice loops coming in from (and
going out to) stations around the world—not to mention routing those
voice connections throughout the building.

There was also the pneumatic tube room. It featured a central switching facility for the plumbing that routed the 4-inch-diameter pneumatic
tube (P-Tube) carriers from room to room. Controllers used these carriers to send papers from one room to the next. The P-Tube system was
copied from those found in large department stores, and now found
almost exclusively in drive-through banks and pharmacies. This communication system became obsolete with the rise of email and the internet.

Around the back and in the corners of the building were freight elevators, staircases, air conditioning rooms, and power supply rooms—and
the occasional janitor's closet. It was fun to wander around the building during quiet times, trying to guess what the humming machines
and boxes behind the many closed doors were really doing. There were
building people you could ask, of course, but the place was big enough
that few knew it all. Flight controllers just needed to know that what
happened on the first floor made everything they did possible. So long
as the many technicians and engineers down there were doing what
they did best, the controllers could concentrate on their spacecraft. Lots
went on to support them—it was many years later that I learned of the
secure room where classified communications could come in and go out.

In truth, there was so little interaction with the "black" (classified for national security) world that most flight controllers didn't need to know the room was even there.

While the first floor operated in anonymity, the second and third floors were the stars of MCC. They were conceived as carbon copies of each other, with the main Flight Control Room (FCR) in the center and staff support rooms surrounding them. In actual practice, the back rooms never matched each other floor for floor. The MPSRs were located in different places. Some very specialized disciplines only had a single support room that served both floors. These were practical arrangements because voice loops and the P-Tube system made it easy to operate from anywhere in the building. Core disciplines generally had dedicated SSRs for each FCR. This allowed them to perform multiple activities at the same time, like a sim for one flight on the second floor and a sim for another on the third.

The original Control Center architecture was centered on the RTCC, which fed data to what today would be called dumb consoles. There were no computation capabilities built into the consoles themselves; they were merely repositories for switches and other input devices, plus lights and screens to show data outputs to the flight controllers. It was figured out early on that some local computational capability was required, so small calculators and computers were added to the back rooms. A few were added to the front room area too. As the Shuttle came along, more complexity was needed, so special terminals (MEDS Terminals) were added to the back rooms so that the controllers could communicate directly with the MOC (Mission Operations Computer) to update limits and constants and the like. In the 1980s, minicomputers came along (this was still before the age of PCs), and these began appearing on consoles in both the back and front rooms.

Computer-savvy flight controllers began writing software that helped them better monitor and command the vehicle. They managed to hook themselves up to data and command streams. It was almost a pirate operation at times, but management knew what was going on and, seeing the benefits, encouraged it. Eventually, enough work had been done to create an entirely new Control Center based on the new technology. The result was a distributed architecture. No big central mainframe was required—data flowed into the building and was distributed to all consoles. The flight control disciplines created their own displays and processing to serve their discipline. Commands worked in the same way but in the opposite direction, heading from console to communication systems, and up to the vehicle.

The distributed processing concept called for a totally different building architecture, so a new MCC wing was added to the old MOW, with new control rooms and new back rooms. Those of us who grew up in the old MCC never felt quite the same in the new building, but we flew the Shuttle there for longer than we did in the old place, and it gave us tremendous capabilities that we didn't have with the original system. The new software also gave us faster and better monitoring, plus the ability to write smart code that enabled fewer flight controllers to do the work. As a result, flight control teams began to shrink. Back rooms that once had three or four people went down to a single individual, with smart and simple artificial intelligence (AI) systems helping to watch for anomalies and recommend solutions.

Regardless of all the technology, the "humanware" always remained the same. They were bright, fairly young individuals with quick minds and the desire to do what was best to advance humanity into space. At the same time, when you spend much of your life in a place like this, and it becomes your second home, a lot goes on that is not directly related to the job of flying the spacecraft. You must communicate with others, you have to store and sort information, you have to eat, you have to drink—and you have to create an environment and a culture that promotes excellence without being grim. Mission Control was not just awe-inspiring, it was a fun place to be, and there was humor along with the square-set jaws, laughter along with the occasional tears. Thirty years of Shuttle flights, built on the legends of the twenty years of space flights that went before, created a lot of new stories. Some of them were mostly true, some of them became altered with time, but all of them are acknowledged canon for those who called themselves flight controllers. What follows is a collection of remembrances of those good times, those interesting times, and those times where things just got a little weird. Life is like that—no matter where you might live it.

Doctor CAPCOM

There are many people who assume that the stress of being a Flight Director must be off the scale when there's a Shuttle in flight. In one way, they're right—there is a lot riding on every decision and how we do the job. But the people selected for the job generally learned to cope with the stress to the point where it really didn't register in our minds for the day-to-day business. You either had the right temperament for the work or you didn't. In other words, if you are given to reacting negatively to stress, then you might not be well suited to the job.

There are, however, times when stress can get to even the most steady among us. It usually hits in subtle ways. Some experience little personality changes. Others have trouble sleeping. In my case, the buildup of stress during a complex mission registered as a tight spot between my shoulder blades, in my upper back. It didn't happen often, but when it did I'd notice a sore spot that bothered me when I moved my shoulders, progressing to a point where it hurt every time I moved my neck. On a bad day, it got to the point where I didn't want to turn my head at all!

On one of my early missions as a Flight Director, I started to get stiff on day one. By day three I was not having a good time—and it was evident to those around me. I could function, I just didn't want to move. My CAPCOM was the legendary space walker and astronaut Story Musgrave…Dr. Story Musgrave, to be precise. Story was a fascinating guy. We usually assumed that he was an alien, visiting from another planet. Overachiever was probably his middle name. He had multiple bachelors degrees, several masters, and was an MD and surgeon. Hired as a science astronaut during the Apollo program, he had been sent to flight school to learn to fly T-38s, and near the end of his career he was the highest-time T-38 pilot in the world.

Story was always into something interesting, always exploring new ideas in science and philosophy. For several years, he carried around a bundle of note cards that he was using to develop an entirely new system of keeping track of information. He was also well known in the Control Center for his oatmeal box. It was one of those round Quaker Oats boxes, and he used it to hold whatever it was that he was eating. As I recall, it was dry, and he ate it with a spoon. It obviously worked to keep him healthy, because at the age of eighty-four he is still incredibly active.

Story walked into the MCC for handover on the third day of this mission. He saw me hunched over the handover notes, obviously not smiling. After we had both taken a look at the notes from our off-going counterparts, we exchanged a few pleasantries as the rest of the team was catching up and getting ready for the handover briefing that I'd conduct on the private AFD (Assistant Flight Director) Conference loop. He asked me how I was doing, noting that I seemed to be looking a bit like a hunchback. Good observation there, Story!

Well, I allowed that my neck and back were really bothering me. I figured that if all went as usual, it'd go away in a few days and that there wasn't much I could do about it until then. "Would you like me to fix it for you?" Story asked.

"Well yeah—if you can do something, that'd be great!" was my response. I've never had any luck though. I usually just let it go away on its own; the muscles would eventually relax.

"No," he said, "we can fix it. Lie down on the floor on your back."

"Hmm...this is going to be interesting," I thought as I found a clear spot on the floor behind the console and lay down. The old carpeting was clean enough, and hey, the lighting was low enough in the room that no one would notice if I got anything on my shirt. Story kneeled down by my head and told me to relax. The next thing I knew, he had put one hand under my chin and the other behind my head. He asked me if I was ready. "Sure," I said, not knowing what to expect. I felt a tug, then a pull—as he dragged me along the floor by my head about 3 feet!

"There now," he said, "Get up and see how that feels." I gently sat up and noticed that the off-going Flight Director was now staring at me, as was the off-going CAPCOM—as was, in fact, just about anyone in the room who was nearby. Everyone was speechless as I slowly turned my head and flexed my shoulders. Sure enough, the pain was reduced to almost nothing, and I had complete freedom of movement! "Wow," I said, "that's amazing! I feel great!"

"All you needed was a little traction," Story said in his small voice. "I just lined up your spine and straightened it out a bit."

I couldn't thank him enough, but we were ready to get started on the handover brief. It wasn't until we got going that I realized just how funny that must have looked to the two full teams in the room—the oncoming CAPCOM dragging the oncoming Flight Director across the floor behind the console. I turned around to my PAO (Public Affairs Officer) and asked if any of that had been caught on camera and gone out to the world. She just smiled and said that she figured it came under the category of medical privacy and had switched to a recorded feature. No need to let the outside world know that we weren't supermen and superwomen after all.

The Coffeepot Abort

The first time I was taken to the Control Center as a co-op student, I felt like I had walked into a cathedral. There was a simulation going on—a rehearsal for the long-delayed STS-1 mission—and I tagged along with one of my officemates who needed to drop something off with a backroom flight controller. I forget the details now, but I remember being amazed at both the unusual nature of the building's hallowed halls and the normalcy of it all. People didn't just do momentous things in this building, many of them spent large portions of their life here.

Among the high-tech consoles filled with blinking lights and CRT screens were book cases filled with mountains of technical documentation. Tables were taken up with primitive computer terminals, and over on the side of each room a station for pneumatic tubes spit out and took in cylinders filled with all sorts of paper that needed to be shuffled around the building. That was the wonderful stuff that let you know you were in the cathedral. The normal part was just as interesting. Mr. Coffee makers sat precariously next to the computer terminals, the trash cans were filled with empty fast food bags, and a wide variety of refrigerators sat hiding behind doorways, almost as if they were trying to hide from critical eyes. Half-emptied packages of paper plates and plastic tableware were scattered about in odd corners, and some rooms even sported early microwave ovens—all signs of habitation by engineers who were spending large portions of their lives dedicated to the religion of spaceflight.

I remember that first visit because I was almost afraid to speak. With everyone wearing headsets and talking softly, it seemed a transgression of some unwritten code to make noise in this place—especially when flight controllers were clearly doing important things. Or at least I assumed they were doing important things—without a headset of my own, it was hard to tell exactly what was going on. As my mentor gave me a quick tour of the building, I quickly lost my bearings. The place was all odd-shaped rooms and consoles strewn about in odd orientations. It didn't help that doors were placed in strange locations, and we moved from room to room without going out into the halls. Only later did I find out that, as programs changed through the years, the Control Center was modified to fit new requirements, but a clean sheet was rarely used. Rather, things were moved a little here, functions changed a little there. Trying to figure out why there was a window in a wall that had been painted over was folly. Someone might remember why they had needed glass there, and then decided it was easier (and cheaper) to paint it rather than construct a whole new partition, but finding that person could be difficult.

I do recall, however, ending up in a hallway with half a dozen doors along one side, and only one on the other. My mentor put his finger to his lips, signaling quiet, and then slowly pushed open the door to reveal "the room." At the time, inexperienced as I was, I only knew this as Mission Control—a place I had been seeing on TV and in books for most of my life. When I was young, I assumed (like most other people) that this room was the entire Control Center. It wasn't until that day of my first tour when I realized it was but a small fragment of the entire building, and that many more people were sprinkled through those tangled back rooms, contributing their expertise to make spaceflight work.

I peered into the room I had seen so many times, not quite believing that I was really there. The cathedral of spaceflight—the nexus of missions past and those to come. With carpeting on the floor and the controllers concentrating on their displays, speaking softly into their microphones, the religious aspect was magnified—it really did feel like I was in church. My mentor looked around and then signaled me to step inside the room. For the first time I stood in the Mission Operations Control Room (MOCR)—a place where I would spend so many years of my working life. We walked across the room in front of the consoles, across the lower level of the stadium seating, and I felt so conspicuous there in the focus of all those eyes. It was only later that I realized that everyone was too busy to notice. In fact I often paid no attention to what was going on down in front of the Trench (the front row of consoles). Sure, controllers were looking at the big screens on the front wall, but anyone walking across the front of the room was below them.

Swinging around the last console, turning right to go out the rear door, we passed the king of all coffeemakers—not a Mr. Coffee, but a commercial unit with two pots to fuel the energy needs of the people in the front room. There may have been a box of donuts or kolaches on a table nearby, or some leftover cake or cookies. This was, after all, a place where people lived—just like in the back rooms. The front room was more formal—ties were always required for men staffing the consoles, and there was a similar level of professional dress for the women—but that didn't change the fact that life goes on, even when you are doing something extraordinary.

That was back in the old Control Center. Later on, I was a Flight Director in that same room, sometime in the early to mid-1990s. Back then, the coffeepot for the FCR was in the little entrance tunnel on the right side of the room. It was located back behind the tiers, which was out of the way but convenient because you didn't have to leave the room to fill up. Now I must admit that I was of a generation that didn't drink a lot of coffee. Caffeinated cola filled that need just fine. I think that part of the reason I never picked up a strong coffee habit was because I never felt like I needed warm beverages in Houston—any time of the year.

Nevertheless, we had quite a few folks who partook of the community coffeepot, and the FCR had a good one—a double burner unit where one side brewed and the other side kept the spare pot warm. Redundancy in all things is important to flight controllers. When you needed to brew a pot, you poured a measured amount of water into an opening on the top and let it do the rest of the work. Unfortunately, the top of the coffeemaker also sported a cooling vent of some sort, and you could see the wiring through the vent if you were tall enough to look at the top directly.

One night during a mission—it was very late, probably after midnight—we were putting the crew to bed and getting ready to start the planning process. All was quiet until suddenly, there came a loud *BBBRRRAAAAAPPPPPP!!* sound from the right rear corner of the room, accompanied by a noticeable flash of light. The noise was followed by some sizzling, and intermittent flashing, as you'd get from an electrical device giving off sparks. The room was quiet no more!

The Ground Control (GC) operator was situated in the left front corner of the room. Concerned about his building, he made a sprint across the front of the room and around the corner. Some rustling and banging around ensued—all of this occurring in a few seconds, of course. The GC emerged from around the back corner with the call, "It's all right, Flight, I think the fire's out!" Sure enough, someone was making coffee and had managed to pour the water into the electrical cooling vent instead of the reservoir opening. Dumping a bunch of water on hot, live electrical wiring is never a good thing, of course, and the results were spectacular. The GC had quickly pulled the unit away from the wall and pulled the plug, solving the immediate problem. But now we had another.

Could a Space Shuttle mission continue without coffee? There were controllers already nervously fingering their coffee cups, wondering what they were going to do on this long night without any java. It wasn't quite an Apollo 13 moment, but there were traces of angst in the eyes of the team! About that time, the crew called down, and CAPCOM told them to stand by a minute because we were sorting out a ground problem. He added that it had to do with critical equipment, and that we were evaluating. (The way he said it made it clear that if this were a *real* emergency, we'd probably be letting the crew know the specifics quickly.) In the meantime, what could we do for them?

After dealing with the crew's question satisfactorily, the crew asked how we were doing with our little problem. By this time we had sorted out some important points, namely that the upstairs FCR was quiet during the mission and nothing was going on there; and, more importantly, it had an identical coffeepot! The GC quickly rounded up some technicians and a four-wheeled cart and dispatched them to retrieve this critically redundant component. So when the crew asked, we were able to inform that after consultation with the flight rules, we were not going to have to bring them home early—that the redundant coffeepot would be in place within minutes, and that we were Go to proceed to the end of the mission. We were quite certain that, come morning, we'd be able to procure a second redundant pot from elsewhere on site, and as a result be back to full capability.

The Phone Under the Floor

When I first showed up in Mission Control in the early 1980s, most of the hardware in and on the consoles was essentially what had been around during the Apollo program of the 1960s and 1970s. It can be hard to remember what the state of the art in electronics was like back then. A personal computer was likely to be an Atari 400 or Commodore 64, or one of those much fancier Apple IIs. Real computers were still mainframes, and punch card programming was still the rage for serious work. Microcomputers took up about a rack worth of space, and seeing images on screens was still a bit new.

In order to make the Mission Control Center run, it took more than computers, of course. Voice communications were essential, and a large portion of the floor space was required just to contain all the patch boards and switching gear. It was easy to forget just how much work and equipment went into allowing you to punch a button and talk to someone on the far side of the globe—or on the moon. The voice system worked amazingly well, especially when you considered that the panels on all the consoles were essentially electromechanical devices with lots of moving parts. These panels—known as keysets—were built by the phone company and were, in fact, connected into the worldwide phone system. You could talk on a NASA communication circuit (known as a loop) or punch a button to get a dial tone and make a call to order a pizza.

Because the system was somehow tied into the phone lines, the equipment was actually provided (and I think owned) by the phone company. Most people alive today don't realize that you didn't use to buy a telephone—you leased it from the phone company. You weren't, in fact, allowed to own your own phone and connect it to the system. The same was true with the government, including NASA's Mission Control.

The phone company had all sorts of odd rules, including one that said that if you had a Public Access Branch Exchange (what we called a keyset), you had to rent a telephone to go along with it. We, of course, used headsets, but NASA was still obligated to have those handsets (telephones) attached to each keyset.

For that reason, if you had lifted a floor tile behind any console that had a keyset installed (and most consoles had two), you would find under the floor, below the wire racks, a standard black telephone—unused, but required by the rules of Ma Bell.

Console on Fire

The old green consoles we used in Mission Control until the mid-1990s were holdovers from the Apollo days. Some upgrades had been performed along the way, mostly because parts for the old things were no longer available. The consoles were amazing for their time, and they eventually became amazing because of their age, but I learned my craft on them. I spent lots of time pushing buttons that clicked when you mashed them, and using the ever-popular digital encoders that you used to dial up channels or display numbers—or anything else you wanted to relay to the computer in the basement.

For real, these consoles were the dumbest of computer terminals. Essentially they were hardwired to the computer(s) on the first floor of Mission Control. You pushed a button to tell the MOC something specific to do, and it responded by sending out hardwired responses to a wide variety of lights and digital displays. The digital displays (the screens) were amazing in their own right, because they were basically black-and-white TVs. They displayed whatever was assigned to the channel that was selected. The MOC was where the digital data came from.

Well, those old black-and-white monitors really were old—right down to the vacuum tubes in their chassis. If you opened the top of the console, you could look at the guts of the display. It was just like the back of an old TV set. Dust and dirt could (and did) get in there, and that led to a fascinating exchange on the MCC comm loops one day when the entire world was watching.

The event was the last major dress rehearsal for the STS-26 Return to Flight mission. Because we hadn't flown anything for two and a half years—and because we were training a lot of people who were new to their positions—we decided to do a full-up, end-to-end simulation, with all the payload control centers, remote sites, and all the major and minor players involved, just as if it were the real thing. To top it off, NASA decided that it would be great to invite the press to watch the whole event, just as if it were a real mission. No pressure, of course.

Things were still somewhat primitive in those days, and it was important to start exactly on time to keep all the players in sync. Therefore, we all got on console earlier than normal. Careful checks were made to ensure everyone was going to be ready to go on time. The press was watching and listening—they were almost embedded in the sim, with representatives from the major networks watching in MCC and in the simulator with the crew. The Flight Director who would be kicking the whole thing off was Gary Coen, Ascent/Entry Flight Director for the mission. With twenty minutes or so to go before it was time to pick up the launch count, Gary called all the players around the world onto the AFD Conference loop to prebrief the event. He talked about how important it was to treat this as a real flight. He expressed his confidence in the teams, and how eager everyone was to prove that they were ready for flight. He went over the essentials of the timeline and covered some last-minute details and changes. All of this was routine prebrief stuff.

Then, as we came up on the end of the brief, he wrapped up with some great words to bring everyone to a high state of readiness and confidence, ending with something on the order of, "So, if everyone is ready to go, let's all head back to our respective loops and get ready to go to run on time. Oh, and GC, CAPCOM's console is on fire..." No excitement, no rise in the pitch or tone of his voice—he didn't want anyone to get excited but... sure, enough, smoke was rising from the CAPCOM's console, right next to Coen. He had said it in such a calm voice that it took a few seconds before everyone realized that he was serious. One of the two GCs in the room responded quickly with a fire extinguisher to vent a little CO_2 into the smoldering remains of one of the CAPCOM's two monitors. There was no delay in getting the sim started—after all, the CAPCOM really didn't *need* to be looking at any data on ascent. The faulty monitor was changed out fairly quickly after the ascent portion of the sim.

It really was just one more step on the road to flight readiness, and it showed just how well the team had been trained by the sim instructors and supervisors. The calm, low-key response was a good indication that this team was ready to fly—even with their consoles on fire.

AFD Conference

"All operators meet me on AFD Conference with an amber please..."

If you were analyzing MCC loop talk, this might be one of the most common phrases used throughout the history of Houston's Mission

Control. Its meaning is simple, but the subtleties are many. AFD Conference was a communications loop that was available to all the flight control positions throughout MCC, but not to anyone outside the building.
Like all communications loops, it was a "party line." Anyone who had
it could listen or talk as necessary. As a general rule, it was treated like
a front room loop—that is, only front room fight controllers talked on
it. On rare occasions, when a front room person had stepped off and the
Flight Director needed to talk to someone in that discipline immediately,
he might ask one of the backroom controllers to come up and give him
an answer on AFD.

Now, most backroom positions also had talk capability on the Flight
Director loop, but there was a reason that AFD existed, and it was
because it didn't go outside the building. The Flight Director loop (or
simply Flight Loop) was the primary execution loop for all operations
once a vehicle was in the air. It was where all important discussions
were held by front room folks and various lead controllers in other centers around the globe. Because it was available to everyone working a
mission—including those outside MCC-Houston—it went *lots* of places,
including squawk boxes in management and leadership offices around
the various space centers, and in Headquarters. It also went to astronaut
family bedrooms and living rooms. When speaking on the Flight loop,
one had to always remember that there were—at one time—about three
thousand external end point terminations. That means there were three
thousand places where it could be heard outside the building.

Having everyone formally fly a mission by referencing the direction
and discussions on a single communications loop is a great idea. It's how
you keep things coordinated. But there are many, many times when you
have something to say to folks just inside the Control Center that really
doesn't need to go outside the four walls of Building 30—and sometimes,
it might be downright embarrassing if they did. And so, the AFD Conference loop was created.

AFD is short for Assistant Flight Director—a position that was created back in the Gemini and Apollo days (it might go back to Mercury, but no one remembers). The Assistant Flight Director handled logistics and special assignments for the Flight Director. It was a person who might very well be training to be a Flight Director, or one of those people we later came to know as Integration Officers—a person responsible for hashing out the details of a plan that had been outlined in broad strokes by the Flight Director. In fact the Assistant Flight Director position later became formally known as the Ops Integration Officer, and at some point the higher-level functions of integration were reabsorbed by the Flight Director while the more detailed ones were absorbed by other positions, many of them going to the Ground Control (GC) position. But back when the AFD position existed, the AFD Conference loop was created. It was a loop where anyone could reach the AFD, and it was the place where the necessary members of the team could hash things out. Because discussions could range through lots of "what-if" scenarios that outsiders might not understand were simply hypothetical, it was felt that the loop needed to be kept within the confines of the building, where everyone sort of knew how ideas developed. For instance, an AFD might be given the assignment to work through a number of contingency scenarios, and if the NASA administrator punched up his squawk box at the wrong time, he might wonder why those crazies in Houston were talking about bringing a mission home early—or working on rescue scenarios. Some things, like sausage making, need to be done behind closed doors.

When the AFD position went away, the loop stayed because it was found to be a great place to hold those "inside baseball" discussions. While anything official needed to be done on the Flight Loop, the AFD loop became a place where the Flight Director tried things out on the team and held preliminary discussions before exposing them to the world. It also was a place where such things as team handovers are conducted. An oncoming team appeared at their consoles at the appointed time—the start of a one-hour handover period. They were generally given the first half hour to come up to speed on what was going on in their discipline, and the mission in general from their discipline's point of view. At the half-hour point, the announcement came that, "All oncoming flight controllers, meet your Flight Director on AFD with an amber." (The color could also be a red or a green, depending where the toggle lights were last left. A light indicated when someone was up on the conference loop, and the idea was to make sure everyone was on the loop.)

Once everyone had come up on the AFD loop, and indicated their presence with a light, the handover briefing would begin. The oncoming Flight Director would go around the room, getting updates, coordinating ongoing work between disciplines, and making sure that everyone knew the plan for the upcoming shift. Since a lot of these discussions really could be incomprehensible by those not deeply involved in the mission, and a casual listener might very well get the wrong idea if they punched it up at the wrong time, the AFD loop was the perfect place for such things.

But handovers are not the only things that went on (and still go on) on the AFD loop. Sometimes, a Flight Director might feel that his team is having a bad day and that they need a pep talk to get them pointed in the right direction. The AFD loop is good for that. Sometimes the Flight Director might feel that his team is having a bad day and needs a dressing down—that is *also* a good use for the AFD loop. Only those inside the building will hear Flight's musings on just how the team is failing, and what they need to do better. In this same vein, the AFD Conference loop is generally used for debriefings—those lay-it-all-on-the-table discussions where you have to check your feelings at the door and take your praise or your lumps. The AFD loop is a great private, safe space to hold those discussions.

Then, of course, there are simply some logistics items that need to be communicated throughout the building but are irrelevant to anyone outside the building. Not only irrelevant, but things you simply don't need them to know.

"Okay, this is Flight—I've got everyone on AFD . . . so CAPCOM and I are going to be buying pizza for the team. I need someone—who's got nothing going on? Ahhh...DPS, you look like you need something to keep you busy. Everyone get your pizza orders in to DPS in the next thirty minutes please—GC, could you put a clock up—call it P Minus or something. When the clock expires, the order goes in. And give me a red when you've completed that..."

Winter Dress Code—in Summer!

Back when MCC was designed, a computer wasn't something you carried in your briefcase; it was something that filled an entire large room. In addition to requiring lots of space, a computer required a lot of power, and most of that power was cast off in the form of waste heat. Lots of computers generate lots of heat, and the Control Center had *lots* of computers. It was therefore realized early on that the building definitely required a heavy-duty air-conditioning system. Even without computers, remember that the Johnson Space Center (JSC) is located in Houston, Texas, which can be simply unbearably hot during the eleven-month-long summer. JSC really didn't have need for a heating plant. The computers could do this job for us.

My first Control Center position was in the Payload MPSR, a back room that was located around the periphery of the building. It held about half a dozen consoles, some racks of computing and communications gear, lots of cabinets and bookshelves, and tables, chairs, refrigerators—all the things a team needed to make a mission work. But this was not the first incarnation of this room. Originally, it had been a full computer support room, with rows and rows of processors, all crunching numbers and turning electricity into heat as a by-product. Because of this, the air-conditioning flow to the room was set quite high.

Oftentimes, to save on electronic controls (that often didn't work), the heating and air-conditioning systems in the buildings at JSC were adjusted mechanically to apportion the flow of cold air to each area—and this room was apparently one of them. Apparently, no one had bothered to tell the facility folks when all those heat-generating computers had been removed and the space was set up for flight controllers. As a result, we had an excess of cooling and a shortage of heat generation, the net result being that when we went to spend a night in the MPSR, thick boots and parkas were often needed to stay comfortable—at least for those controllers who hailed from warm climes. Being from Minnesota, with a reputation to defend, I toughed it out without a jacket and wore a short sleeve shirt. But I can reveal the truth these forty years later. It was freezing in there!

I was glad that my promotion to the front room came quickly. That particular room eventually emptied out, and it was rarely used until that entire side of the building was renovated, including the air-conditioning.

P-Tubes

Way back before the internet, far back in time before the age of the personal computer, the MCC (Building 30) was built. Getting written material from one place to another was a logistical challenge. This was solved by calling on an age-old technology everyone was familiar with from large department stores and bank drive-through windows—the pneumatic tube.

The P-Tube system was part of the initial building design, and its pipes ran everywhere. No matter where you were (except in the restrooms), a P-Tube station was nearby. You could send a carrier tube from any station to any other station by dialing in the destination's code number on a dial (in the back rooms) or by pushing the correct button on the fancier panels (in the front rooms). In the back room, you put the carrier tube in the system, closed the hatch, and hit Send. In the front room, seen by millions of people on TV, you pressed the correct button, and the hatch magically opened. Then you dropped the carrier tube in the open slot, the hatch closed, and off the material went with a whoosh.

What many did not know is that there were actually two systems—and if you were sending from a station on one system to a station on the other, it took manual intervention in the P-Tube room to swap the incoming carrier tubes to their outgoing destination. Operators were there 24/7 to make sure that P-Tubes got to their destination, regardless of snow, sleet, or gloom of night, I guess. During busy times, visiting the room was fun, because you could watch the guys grabbing tubes, checking addresses, and sending them off to where they needed to go. The basic system worked great—and it was efficient. The P-Tube system was not duplicated in the new Control Center, because everyone (by that time) had a PC and an internet connection. Paper was replaced by electronic files, and transfers were done at the speed of light. That is, when it worked. In the early days of the new system, the internet often failed and so message transfer was accomplished by Sneaker Net—someone had to run a printout from one room (or floor) to another.

Now the P-Tube system had strict rules about what could be sent. It was two ounces or less of paper, per carrier—and nothing else. Paper only. Of course, flight controllers were engineers, and engineers love a challenge. They especially love to challenge limitations. As a result, many items other than just paper were sent through the P-Tube system. Now, of course, I never sent anything other than paper you understand, but there are stories...

Way back in the day, someone in the front room needed some pencils, so he asked his back room to bring some up. For some reason, the backroom operators decided that was too much work, and besides, there was a quicker way! So they loaded up a P-Tube carrier and hit the dispatch button. No one knows why, but the carrier door opened up en route and pencils spilled into the tubes. Legend has it that technicians were picking splinters of wood out of the system for years afterward.

Then there was the time that the legendary Ed Fendell, Captain Video himself, was working the Instrumentation and Communications Officers console and asked his back room for a ham and Swiss sandwich. His back room replied that they'd send it via P-Tube. I wasn't there at the time—but I have read the transcript of the ensuing conversation on the various communication loops. Apparently, annoyed that his sandwich hadn't shown up, Fendell asked his back room if they had sent it. Assuring him that it had gone, Ed began hunting for the sandwich by talking to other rooms throughout the Control Center. This meant that he had to punch up the appropriate comm loop for whomever might be by the various P-Tube stations throughout the building, and giving them a call.

What Fendell didn't know is that his back room was following him on the loops and recording all the conversations. Each time Fendell called someone new, he had to explain about the sandwich, and then ask if it had shown up there. What he didn't know was that it was all a setup—the sandwich had never actually been sent! The "Great Ham and Swiss Hunt" went down in the historical lore of MCC, proving that you have to be careful when talking on a recorded communications system.

Ya Gotta Eat!

Food has always played a significant part in MCC operations. The onsite cafeterias apparently supported a snack bar in the Control Center back in the early days, but this did not survive the down years between Apollo-Soyuz and the Shuttle era. What we were left with was vending machines. While you could make it through a shift with a candy bar and a soda if you had to, you could almost always find food (if you looked hard enough) that had been brought in by flight controllers to share with their teammates. Morning simulations or shifts that started at dawn, Houston time, were always a good time to look around for kolaches. Strongly reminiscent of pigs in a blanket, the kolache was food that you could pick up at a local bakery by the dozen for a few bucks and share with the team. It became almost traditional that if you were being certified on a particular sim, you brought in kolaches—or donuts—for the team to enjoy in exchange for them suffering through a script that was heavy on items for the discipline being tested and light on everything else.

Some disciplines were well known for particular "feeds" during a mission. Some might have a pasta night, or bring in several pans of lasagna. Of course, cakes, pastries, and cookies were usually in abundance as well. A person could gain a lot of weight wandering around the Control

Center during a mission if they weren't careful. Food was less plentiful during simulations, but you could find it if you knew where to look. There were also a couple of fast food joints just outside the front gate of the space center, though it was considered bad form to sneak out to pick up lunch if you had a gap in your work—unless, of course, you offered to pick up something for those around you.

It was no secret around Mission Operations that I was a fan of pizza. I took it on myself to always try and have a pizza night for my team during a mission. This was hard to do if we went on console before midnight and got off at the crack of dawn, but if the timing was right we tried to pick a day near the end of the mission, when things had gone well and there wasn't a lot going on. I'd usually have a cohost (or coconspirator) in my CAPCOM, and they'd offer to split the bill for pizzas for the entire team. We'd find one of the disciplines that didn't have a lot going on and ask if they could be a clearinghouse for orders—then give everyone a deadline for providing their needs to the coordinator. We'd ask for status light colors and set timers counting down to the pizza order (just to make sure everyone knew when they had to have their requests in). When the deadline came, the order went in. At one time, there were certain pizza restaurants that had a delivery driver with a JSC site badge. Those drivers could bring the pizza all the way to the Control Center. Other times (particularly post-9/11), we had to send someone out to the gate. It usually took about two dozen pizzas to feed a flight control team, so it was no small task to get it all delivered up to the front room. If we ordered an extra one for the gate guards, things seemed to go smoother...

When the pizza arrived, we'd make sure that the front room cameras were not going out to the world—we'd have the PAO go to a playback or a recorded feature—as we didn't really need to advertise for any particular pizza brand on international TV...on a US Government feed. Executed properly, we'd have everyone fed and the boxes cleaned up in an hour, with the leftovers out of sight by the coffeepot for continued snacking. And it should be noted that while pizza is probably not on the "health food" list, I never saw our flight surgeons (or their support teams) turn down free pizza.

PAO's Atlas

Although the Shuttle flight control teams occupied a number of different front rooms over the decades of the Shuttle program, one constant followed us wherever we went—a large and detailed Earth atlas, kept at the Public Affairs Officer (PAO) console. I don't remember the date on the cover, but it clearly went back to before the Shuttle. PAO kept it handy in case they needed to figure out what they were seeing on downlink TV when the crew (or ground) was panning down to show Earth. It was the final arbiter of discussions over "Which lake is that?" or "Is that the Danube or the Rhine?" As useful as it was for things like that, it needed to be carefully used when looking up the names or borders of countries—especially small African nations. They had a tendency to change. And even though the atlas went on a shelf when Google Earth came along, it was still referred to now and again by those who loved the old ways. Sometimes, you just want to have a real book in your hands.

Visit from "Space"

It was not uncommon—in fact, it was probably the rule rather than the exception—that any Very Important Person visiting the Johnson Space Center ended up in the front room for at least part of their tour. I have hosted princes, kings, and presidents. I have chatted with singers and sports figures. And I have shown off what we do to teachers and their young students—the most rewarding tours of all. In general, most of these folks were friendly and engaging, many of them memorable—but one visitor, in particular, was simply a delight—and special in a nostalgic way.

It was late one afternoon or early in the evening on a Shuttle mission, with the crew already in bed, when I was told that we had a visitor coming, a friend of the Administrator's. When I heard the name, it wouldn't have made any difference to me if she was an enemy of the Administrator's—actress June Lockhart had played such a special role on television during my formative years in the 1960s. *Lost in Space* was a three-year-long science fiction show that, in retrospect, was more than campy and had very little science. But the tale of a family, their robot, and their spaceship running around the galaxy trying to find their way home was captivating to a young boy enamored by all things that flew. Ms. Lockhart played the mother of the family, Maureen Robinson, and was always the voice of reason and steadiness—a true Midwestern mother in my book.

Ms. Lockhart's visit was fun because she really was a delight—a genuinely nice person who seemed to really be impressed by what we were doing. I spent some time explaining the control room and what the Shuttle was doing in space—how we were watching over the crew and the vehicle while they slept. She spent quite a bit of time with us, and I am surprised that I don't remember that she had any handlers with her—she was just comfortable being with us by herself. At one point, we were sitting together by the Flight Director console, and I told her just how much I had enjoyed the *Lost in Space* series when I was a kid. She sort of laughed and said something about just how silly it was in retrospect—and I must admit, the plots got pretty goofy when they ran out of the obvious space-genre action scripts. She seemed proud of the work but didn't think it was anything special.

At that point, I paused and said, "I understand what you're saying. And in light of today's more sophisticated take on science fiction, and television, and given what they can do with special effects, it does seem like it was a pretty low-budget, simple show. But if I were to ask the team here, right now, how many of them watched *Lost in Space*, either during its first run or later on [in reruns], and then ask how many were inspired to be a part of the space program because of it, I am willing to bet that 80 percent of the men and women here in the room would raise their hands. You see, early learning is strong learning, and we grew up with the idea that people could live and work in space because we saw TV shows like yours. Don't ever feel bad about what you did then—it was inspirational, and paid dividends far beyond what you might have ever thought."

I think that I might have seen a glint of a tear in her eyes as she thanked me for that perspective. It truly is hard to know just how our actions will inspire others, and what effects we'll have on the generation that comes after us. And in the end, that is really what the space program is all about—inspiration. June Lockhart inspired a generation to get into the space business, and that generation will inspire others to do the same thing. It all rolls into the future, and while we can't tell exactly where our efforts at NASA will lead, we bet that sometime in the future those efforts will lead the way for young people to leap beyond our mere beginnings.

Execute Package Humor

The Execute Package is the crew's morning newspaper—it tells them what they need to know about the coming day and includes timeline revisions, procedural changes, and various news items such as consumables quantities, the weather at emergency landing sites, and the like. It also oftentimes contains a little actual news from the world (submitted by the Public Affairs Officer—something to keep them awake on the overnight shifts). The Execute Package is put together by the planning team, based on inputs from the previous day's Orbit 1 and Orbit 2 teams, plus necessary materials submitted from around the mission ecosystem.

It generally took the planning team most of their shift to pull together the inputs, format them, check them for accuracy, edit the resulting pages, and then review everything in time to have it uplinked. The Execute Package could be relatively short, or it could be a whole "special edition" that included massive rewrites of procedures. The first thing the planning shift Flight Director did when coming on console was to get an idea just how big of a job the team had ahead of them and adjust the pace of the shift accordingly.

Now for much of the Shuttle program, the Execute Package was distributed to the team as paper copies and slipped into the console inboxes about the time the crew woke up. The crew got theirs via an uplinked teleprinter message early on, then via a printer connected into the computers later, and finally, by direct uplink to the laptop computer network with no attempt by the ground to figure out what the crew wanted printed and what they did not. In those later years, the flight control team went paperless as well, with the package available on the Control Center computers for easy access. Some printed certain pages, but in general we worked from our screens.

Notwithstanding the actual form of delivery, the Execute Package was still referred to as a package, with a cover sheet that listed all the messages contained in that day's edition. And it became traditional—no one I ever talked to could trace down the exact time when it started—to put a cartoon on the cover of the Execute Package to give just a little levity to the morning read. The cartoon usually played off something going on in the mission: either making a little fun of a Control Center discipline that had something going on, or of a payload that was either having trouble or was performing in an outstanding way, or oftentimes having a little fun with the crew. The funny thing about that was, until we entered the digital uplink age, the crew never saw the cartoons until they got home—we didn't have picture uplink capability! It was always a little unfair to pick on someone who didn't know it—kind of like putting a "kick me" sign on their back—but we judged who would take it in stride and who might get angry...and it was well known who would react in which way.

Flight Directors were always given a choice of Execute Package cartoons, just to make sure that the Flight Activities Officer's (FAO) back room (who were responsible for generating the cartoons) didn't go beyond the pale of reasonable taste, or make a faux pas because they weren't aware of political, social, or personality characteristics that might be going on behind the scenes. Being a fan of edgy (but not over the edge) humor myself, I always told my FAOs that if I didn't reject at least three or four cartoons during the mission, they weren't dancing close enough to the cliff. Somewhere in all my boxes of memorabilia, I probably have a folder of rejected cartoons. I'll have to look for them someday.

One of my favorite cartoons poking fun at me and my fellow Flight

Directors was a simple one. It depicted a person standing in front of the coffeepot, arms folded, with a badge that said "Flight" on it. The person was staring at the machine and barking out, "Make Coffee!!" It was a commentary on the fact that Flight Directors never actually *do* anything themselves—they direct others to make things happen.

If the FAOs didn't have a reasonable artist in their back room for a mission, they would often pull a cartoon out of one of a large stack of cartoon books they kept (*The Far Side, Calvin and Hobbes, Peanuts, Bloom County*, and others were the likely donors) and alter the text to make them appropriate to the situation. Another favorite was a *Bloom County* cartoon featuring Bill the Cat in which the other characters were preparing him for a press conference. Bill was wearing a football helmet and pads. Bill had a name tag that read "Flight," and as one character asked a question in the nature of "So, Flight, can you comment on the report that the crew reported seeing a UFO?," another character swung a baseball bat and clobbered him right on the helmet. Yeah, I have seen press events like that...

It was, of course, a great honor to be lampooned on the cover of the Execute Package—and an event to be cherished in your Control Center career.

The TLI Memorial PAD

A PAD is a form used to transmit information by voice from the ground team to the crew in the most efficient manner possible. There are many different types of PADs, but they work on the principle that if both the sender and the receiver have the exact same form, then all you have to do is read up the information without having to read up explanatory things like what each piece of information means. A Burn PAD, for instance, contains information that the crew needs to perform an orbital maneuver. It consists of the vehicle attitude, the amount of thrust in each direction, special notes, and so forth. So, instead of the CAPCOM having to read up "Roll equals 045, pitch equals 137, yaw equals 067," he looks at his PAD that has the information with the exact roll, pitch, yaw order.

Knowing that the crew has the same PAD, he tells the crew which line they are starting on and simply says, "045, 137, 067." If you have a lot of information to get through, this process can really speed things up.

The Shuttle had a standard maneuver PAD for exactly this purpose, and even when we could uplink the information later on, we still read up the critical data in case an uplink went bad and a wrong number went up digitally. It was a cross-check on our own work. The standard Orbit Maneuver PAD could be used for anything, really—big burns, small burns, altitude corrections, plain changes—whatever was necessary.

So it came to pass one late night in July 2009 that I was working the overnight shift on STS-127. They had launched on July 15, one day earlier than forty years since Apollo 11 left for the moon. There was much discussion about this historic event, of course, and as I recall, STS-127 was going extremely well—there were few changes for the planning team to work, and our biggest problem was keeping everyone awake. I hadn't heard anything from the Trench for a while, so I pulsed my FDO. "Hey FIDO, have you got any idea how much propellant it would take to send the Shuttle to the moon?" It was a physics question that involved a little engineering to account for the characteristic of the rocket engines being used, and I figured that such a question might keep his team busy for a while.

"I don't know, Flight—but I bet we can find out! We'll get back to you." That was my FDO, either grateful for something to do or just humoring the old man's odd request. I let them go chew on it for a while and decided to bother someone else.

In a little while, I got a call back from my FDO. "Flight, FIDO, we've been looking into your question regarding a Trans-Lunar Injection (TLI) burn for the Shuttle and, believe it or not, we still have targeting software that can solve that kind of problem. We blew the dust off it, and at the Orbiter's current weight, it looks like it would take about 240,000 pounds of prop to get us on a free-return trajectory around the back side of the moon. Of course, we'd need some more to keep us in lunar orbit and then to get us headed home. Would you like us to work up a Burn PAD for the TLI? We can do that!"

Of course, they were just needling me back with the Burn PAD. But I figured that if it kept them busy, it would be fun to put the classic Apollo maneuver into Shuttle terms. "Sure FIDO—let's see the PAD!"

Now you have to realize that a fully loaded Orbiter, with payload and consumables, weighed in at around 230,000 pounds when it got to orbit—so the weight of fuel to send the thing to the moon was going to roughly equal the entire all-up weight of the bird—clearly impossible. But it was fun to have that idea rolling around on our minds.

Sure enough, in about an hour, I had a nice Burn PAD to add to the crew's morning mail—the "STS-127/Apollo 11 Memorial Burn PAD." It had a time of ignition, attitude, targets—the whole deal. And knowing the precision and integrity with which our Trench guys worked, I trusted that they had actually run it through the computers and not just made it up. Now there were practical problems, of course—the burn duration with the Shuttle's OMS engines was going to be more than a single revolution of Earth. That was not only impractical from an engineering standpoint, it was going to affect the orbital mechanics as well, because burns were usually targeted to a point in time. But it was fun nevertheless.

We did, of course, put a note on the top of the PAD, **"DO NOT EXECUTE WITHOUT AN EXPLICIT GO FROM HOUSTON,"** just to make sure that we didn't joke ourselves into a sticky situation. I didn't want to be the one who had to explain to management why we ran the Orbiter out of gas trying to get it to the moon...

Wake-Up Music

Wake-up music has been a tradition in the NASA human spaceflight program since astronauts were allowed to go to sleep on orbit—probably on the long-duration Gemini flights. Someone got the idea to wake up the crew with a bit of music played over the air-to-ground channel, and astronauts all being pilots (at the time), and pilots being pilots, the music chosen was often the worst tonal atrocities that could be found—often country and western twangs that could annoy just about anyone. The wake-up music was chosen as a joke by the (astronaut) CAPCOMs to annoy their brethren in space, and was a good source of levity within the team.

No one is sure when it actually became standard procedure to wake up *every* crew on *every* morning with music, but it was certainly a standard by the time the Shuttle came along. The problem was (and is) that coming up with songs every morning can get tough when you've done it over, and over, and over, and over... It's easy to get stale.

By the time I was selected as a Flight Director, it was standard procedure for the Planning Shift CAPCOM to choose the music and make sure that it was ready to play. Since the CAPCOM and Flight Director worked side by side, it was not uncommon for the Flight Director to get involved with the selection—to the relief of CAPCOMs who were starting to run out of ideas. Eventually, the CAPCOMs solved this problem by turning it over to the crews themselves and letting them pick the music they wanted to hear. It was the death of the old days, when the idea was to bond the flight control team and the crew by annoying each other (it was an odd way to bond, but it seemed to work). As Shuttle flights became more routine, it became harder and harder to find annoying music, so this change was probably inevitable.

We still found times to be inspirational and unique, however. There was one early morning when we received word that Charles Schulz, the noted cartoonist and creator of Charlie Brown, Snoopy, and the entire Peanuts gang, had passed away overnight. Schulz had wormed his way into the hearts of NASA's human spaceflight team long before, back in the 1960s. His legacy survives today in the most meaningful award a space worker can receive in recognition of sustained high performance—the Silver Snoopy. The award can only be earned once, and not by a manager. It is presented by an astronaut, and it rewards those who make human spaceflight safer and more successful. Schulz was an icon, and his passing saddened us all.

I recall that we heard of his death just a couple of hours before a crew was scheduled to wake up, and while I have no idea what music we were intending to play that morning, I knew what I wanted to do. I asked the team if anyone might possibly have a recording of the Peanuts music—music composed and performed by Vince Guaraldi for the first TV special about the Peanuts gang. True to the can-do nature of our teams, someone piped up that they had a CD back in their office and had sent someone to get it. We quickly had the comm techs in the basement get it cued up, and we woke the crew with the sounds of the soft jazz piano that all of us had grown up with at Christmastime. It was proof that you didn't need to annoy someone to make wake-up music meaningful.

Chapter 10

Return to Flight

It's never one thing that causes a catastrophic failure—or more accurately, it can be one little thing that doesn't come into play until a whole chain of events happen that make for a catastrophic failure. In modern mishap investigation, this is called the error chain. If we are to understand the cause of a major failure, then we have to follow the error chain from the small individual failure (the proximate cause) to the root cause—the thing that caused the little failure to become a catastrophe. If, for example, you conclude your investigation by saying that a loose bolt took down an airliner, then you really haven't solved anything. If you figure out mechanically why the bolt was loose, then you are closer to the answer. But to really figure out how to prevent the mishap from happening again, you need to keep asking why until you find the root cause for the loose bolt. Usually, that comes down to a human error that can be prevented by some measure. That is accident investigation in a nutshell; and while it sounds easy, it can take a very long time if the causes of the mishap were complex.

We lost two Space Shuttles, and their crews, in the course of the program. *Challenger* went down on ascent in 1986, and the *Columbia* was lost during entry in 2003. Both mishaps were traced to mechanical failures fairly quickly. I recall first hearing that a leak in a solid rocket booster was the proximate cause of the *Challenger* loss within hours of the failure. The specific cause of the *Columbia* breakup took a few days longer, but it was suspected very quickly to be a failure in the leading-edge thermal protection caused by a piece of foam breaking off the External Tank on ascent. While it was good to know both of these things, neither of these facts, taken alone, were sufficient to tell us what really happened so that we could find ways to prevent them from happening again.

In both cases, it took nearly two and a half years after each occurrence to return the Shuttle to flight. In neither case was any of that time wasted. While the engineering community invested a huge amount of time and energy into fixing the physical causes of each mishap, the operational and management communities spent an equal amount of energy working toward finding the root causes—and those causes were deep and complicated. People problems usually are.

On the morning of January 28, 1986, I was standing in the lobby of JSC's Building 4, the office building that housed system flight controllers, the training organization, and the Astronaut Office. I was on my way to a meeting to discuss procedures for the upcoming Astro-1 mission (an astronomical science mission that used the Spacelab IPS for which I was responsible), and I stopped in front of the television, along with several dozen others, to watch the final few minutes of the countdown for STS-51F. I had paid some attention to the launch process, but I hadn't concentrated on it—I was a young flight controller with the next mission to think about. But when I saw those twin plumes of exhaust as the Solid Rocket Boosters (SRBs) took off on their own, leaving an expanding cloud of gas and debris where the Orbiter and External Tank had just been, I realized quickly that we weren't going to be launching another Shuttle anytime soon—much less Astro-1. Our lives changed that day—and the program was never going to be the same.

We'd lost a crew, and at the same time, a vehicle. The vehicle was eventually replaced, but those seven lives could never be brought back. We all knew that this was more than a failure of technology. We needed to figure out how we got to the place that had let us think that it was okay to launch when we didn't really know how cold temperatures could affect the O-rings that kept the hot gases inside the SRBs. There was a serious flaw in the decision-making process, and it had to be solved—as well as anything else in our operational processes and rules that were affected by the same culture.

It was culture that got us in to trouble—culture, success, and the pressure to perform. We fell into the trap of success. With many diverse and complex missions launched in a short period of time, we felt that we knew what we were doing. We felt comfortable because nothing terrible had happened. But what we didn't realize was that the terrible thing just hadn't happened...yet. Simply put, we had lulled ourselves into thinking that since we had gotten away with it before, we could get away with it again.

I like to point out to people that this is like playing Russian roulette. In this macabre "game," you place one bullet into a revolver and spin the chamber. Assuming it is a six-shooter, you now have a one-in-six chance that a live round is lined up with the firing pin and barrel. You point the barrel at your head, pull the trigger, and see what happens. If it doesn't go off, you managed to win—you had five out of six chances of doing so. Now the smart person would have never played the game at all. But let's say you had no choice and had to play. Say the live round didn't line up when you pulled the trigger and you were safe. If you were smart, you would simply put the gun down and walk away, saying that you won and would never play again.

But what happens if you are not thinking it through? What happens if you think, "Well that wasn't so bad—I guess I should try it again!" Well now you are on your way to disaster because every time you pull the trigger without firing the round, your chances of getting killed on the next trigger pull go up—not down. Anyone who survives on luck and wants to live a long time should take that luck and get themselves out of the game of chance. People in the flying business—either atmospheric or in space—are taking enough chances as it is, even when they are doing everything right. Relying on luck is simply not an acceptable way to fly.

But our culture at NASA had developed into one in which we couldn't fail—that we couldn't allow ourselves to fail. That is a good thing if you are talking about actual in-flight operations: we would not allow ourselves to fail to bring our crew back. But when we extended that to launch decisions, which really are decisions on making a mission success-ful (you can't have a successful flight if it doesn't launch), that is where we got into trouble. Flight Directors are responsible for crew safety *and* mission success—not one or the other. That means we have to find solu-tions that satisfy both goals. Oftentimes, you must cancel a launch so that you have an opportunity to launch the mission another day. And by not launching when conditions aren't right, you ensure crew safety as well. Never launching means that the crew will be safe (or at least won't be lost in space—they could still get hit by a bus), but you never achieve mission success. Launching with the maximum opportunity for a safe and successful outcome is the key, and that got lost in our culture at the time of the *Challenger* crash.

But to a newly minted front room flight controller, all of that was beyond my ability or skill to affect. For me and my peers, the downtime after *Challenger* was filled with day after day, week after week, and month after month of detailed work on flight rules, procedures, and training. After a couple of years of continuous Shuttle flights, with one following the other with some-times less than a month in between, we needed to be able to stand back, take a breather, and fix some of the things that we knew were less than per-fect. Even though there had seemed to be an endless amount of time to get things "right" before the first Shuttle flew, until you get a flight program going, you don't know what you don't know. As a result, you might gener-ate a lot of "stuff," but it might not be the right stuff. Put another way, we didn't have the resources to keep improving our operations and operational products and keep flying at the same time. The downtime after the crash allowed us to catch up.

Shortly after the *Challenger* accident, management and I decided it was time for me to move from the Payload Support section into something more mainstream. I then transferred to the Mechanical Systems Section to support Orbiter systems (and a whole lot of other things). As such, I first had to learn a lot about the systems as they were. Then I quickly became involved in modifications that were being made, primarily to the landing and deceleration systems for the vehicle. These included the tires, wheels, and brakes to start. Then there was the nosewheel steering system, which had never worked properly. It was therefore being upgraded to eliminate some single-point failures. These single-point failures were places in the design where if one problem occurs, you have no backup, no recourse—things could be fatal immediately, with a loss of control on the runway. Then we had the challenge of coming up with a new way to relieve the (undersized) brakes of their stopping task, as well as relieving loads on the nosewheel. The answer was a drag chute—and that required a huge amount of operational development as well as design work.

Every group within Mission Operations had similar improvements going on—upgrades to computers, changes to various valves and plumbing configurations, and countless wiring changes being done throughout the Orbiter. Every team was taking advantage of the break to fix all the things that needed fixing but hadn't had time to work on. This by no means meant the Orbiter was a lemon or had a poor design; it just meant that this highly experimental, developmental spacecraft had gone straight from its first test flights and into "routine" operations without a chance to fix all the things we learned about during the orbital flight tests. It was tragic that it took a fatal accident to give us the chance to make these upgrades. But the time was put to good use.

In addition to all the landing and deceleration changes that we were following in my group, we were also being made responsible for the new Launch Escape System—essentially pressure suits, parachutes, and an escape path (a jettisonable hatch and a slide pole to make sure you cleared the wing when bailing out). The escape system involved all new hardware, and that meant all new procedures and rules. There were also upgrades to the Auxiliary Power Units (APUs) and hydraulics that we were responsible for. This work followed and incorporated the upgrades that were made to the procedures and rules. To say that our group was any busier than any other Systems Division group would be to insult those others—everyone had things that were being worked. It was probably the busiest time I ever saw organization-wide in my years in the program.

In addition to all the system changes being worked, Mission Operations took the opportunity of the downtime to rework *all* the flight rules. They were reviewed one by one, and we made changes where we thought they were needed. The rationale behind every single one of them was documented. The Flight Rules book more than doubled in size as we documented the reasons behind all the rules. Everything was then incorporated into a single Rules and Rationales document that eventually grew too big for even a 6-inch, three-hole binder. In parallel with the rules reviews, we were also going through every procedure—nominal and off-nominal—with a fine-toothed comb to make sure there were no trap doors. We needed to be sure that all the physical changes to the systems had corresponding updates made to the operating and trouble-shooting procedures.

In my own little piece of this world, I was not only supporting engineering changes to many of our mechanical systems, I was also learning a new flight control job: that of the Mechanical Maintenance Arm and Crew Systems Ascent/Entry flight controller. My previous experience in MCC (Mission Control Center) was spent operating and managing support systems for payloads, so I was experienced in the on-orbit environment and straddling the gap between the pure Orbiter operations and systems, and those of the payload world. My work started after Main Engine Cutoff. But being an airplane guy at heart, I was ready to soak up the dynamic phases of flight—ascent and entry—like a sponge. I spent day after day in the Control Center, either simming myself or watching others sim. Shuttle ascents were complex and happened very fast. There was no time for long discussions on potential solutions to a problem. You had to know your response instantly—and it had to be right. I enjoyed the high-pressure ascent time frame, of course. But my real love was for entries—the act of flying a winged vehicle through the atmosphere to a wing-supported landing. I have always said that I love landing airplanes. I certainly love flying them up and away, but landings require the delicate finesse of bringing the bird back to an exact point on the surface with a gentle descent rate so that nothing is damaged. It makes no difference if we're talking a Piper Cub or a Space Shuttle—landing requires skill and confidence in a pilot.

So I was quite happy to be thrown into the landing/deceleration world where we were constantly looking for better and safer ways to bring every Shuttle landing to a successful conclusion. At the time of the *Challenger* accident, there had been several instances of burned up brakes and even a blown tire or two. Everything in the landing world was slightly undersized because they were specified at a time when the Orbiter's weight was supposed to be lower. All aircraft gain weight in the design process, and it is important to recognize this when sizing components. Because weight is such a critical factor in a space launch vehicle, the tires and brakes for the Shuttle never really grew as the Orbiter did, so they were very marginal. This was why we looked at adding a drag chute to get the vehicle out of the high-speed regime on the runway as quickly as possible. As a result, the brakes needed to remove less energy. Surprisingly enough, there was a drag chute in the early conceptual designs for the Shuttle, but it was deleted in a weight savings exercise sometime in the mid-1970s. Adding it back required us to do a great deal of design testing as well as procedure and rule development so that we knew how to use it to best advantage.

Drag chute operational development took place to a large extent in some complex simulators, namely the Shuttle engineering simulator at the Johnson Space Center and the Vertical Motion Simulator at the Ames Research Center in California. Hundreds of pilots, both astronauts and engineers, flew thousands of runs in the big six-axis simulator at Ames to perfect ways of landing that gave the most margin—especially with blown tires, locked-up brakes, an unexpected drag chute failure, or failures in the nosewheel steering. All of this flying reflected back into procedures and rules, which then poured back into the training program for both pilots and flight controllers. It was frankly a fun time to be in the program. We were learning every day, and drinking from a fire hose is exciting when you're young.

While all of this was going on in the trenches, upper management was apparently going through its own soul searching. They reorganized things to make sure strategic decisions were made in the open, and that there were opportunities for objections from any quarter. The problem, of course, is one that any large, technical organization faces: it is easy to *say* that if anyone has a problem, they should say so and operations will be brought to a stop until their concerns are satisfied. It is another thing to actually make that happen. With literally hundreds of thousands of engineers involved in the Shuttle program, it would be surprising if you didn't have at least one individual who wasn't comfortable at any particular time. In fact, there are always small groups who are concerned about every launch and would like to see it stopped until more work has been done.

The issue is one of risk acceptance. Some people accept too much risk and take chances when they shouldn't. Taking too big of risks often leads to failure, which can also mean loss of life. Others are too risk averse and will try never to accept any risk at all, and as a result they never do anything. At least they are safe, but their mission is a failure. The trick is to be somewhere in between, maximizing safety while at the same time maximizing the chances of mission success. But a smart team listens to the naysayers, because oftentimes that lonely voice crying out in the wilderness is correct. You ignore them at your peril, because if you fail and kill someone, then it looks like you were being reckless by disregarding the warnings of someone who knew better. It puts leaders and management in a very tough spot. Folks in leadership roles understand that no one knows everything, and they also understand that sometimes the folks who are being more conservative don't have the whole picture in mind. Other times, the folks who are pushing to go ahead don't have all the facts. What is important is to make sure that everyone feels like they have been heard, and that their position has been weighed in the final decision. At least this was how I always saw it—so long as I felt my position was heard, understood, and considered, then I didn't feel so bad if the decision went in a different direction. You simply don't win every position that you take.

I like to tell a story that was told to me by my Apollo colleagues about what they called "Burning Rocks." Apparently, in the push to go to the moon, one particular scientist was concerned about the chemical composition of the lunar surface and its potential for interacting with the exhaust plume of the lunar module. He felt that there was a danger, but he kept it to himself because everyone else was pushing to go, go, go. Finally, he couldn't contain his fears any longer, and just before the crew was going to power up the descent engine on the lunar module to start their actual drop to the surface, he spoke up and said that he was afraid that when the exhaust hit the surface, the moon might spontaneously combust in a cataclysmic event. This was, of course, after the crew had launched, transited to the moon, and put themselves in lunar orbit. The Flight Director was presented with this information just before the Go or No Go decision to begin the descent, and his reaction was essentially, "Well, what the heck do you want me to do with this?!"

There are times and places for considered and supported arguments—and that time is before you have committed to the dangers of spaceflight. Once you are hurtling along at high velocities, you have to accept that you have already taken great risks to get there, so maybe you'll have to take a few more risks to get the job done. However...if you have new information, something that you never had before, then it must be presented and thrown into the mix. Sometimes, you might just save the day.

STS-26 launched on September 29, 1988, more than two and a half years after the loss of the *Challenger*. It carried one of the best-trained crews in Shuttle history, and it was monitored and controlled by probably the best-trained flight control teams in history. The flight controllers never stopped training during the downtime, and I personally went from one orbit discipline to being a front room member of the Ascent/Entry team in that time. It was a marvelous couple of years, despite the tragedy that started it off. The training teams put us through the wringer countless times, yet we all knew that if we were once again faced with a complete catastrophe like that which befell the *Challenger* and her crew, we would still be nothing but spectators at close range, helpless to affect the outcome in a meaningful way. But we trained for the thousands of cases that we *could* handle, and we did our best to ensure that the engineering teams had done their best to preclude the worst cases. Moreover, we were fortunate to be flying a much better, more capable, and measurably safer vehicle.

Shortly after the main engines cut off and the *Discovery* was in orbit, Lee Briscoe, the Weather Flight Director for the launch (he assisted the Ascent Flight Director by assimilating and organizing all the incoming weather information), walked around the room. He leaned over each console and shook the hands of every flight controller, congratulating us on the work we had done to get to that point. Yes, we still had an entire mission to fly, followed by the entry—but the first step was complete. America was back in orbit, and we all knew that we had done a good job with the time we had been given to make the program safer.

The Shuttle flew safely for years after that day, of course. Our flight rate never again came close to what we had done the year before the loss of the *Challenger*, and obviously we were nowhere close to the absurd projections of the 1970s that said we'd be flying sixty Shuttle missions each year. The system simply could not accommodate that. Gone were the commercial satellite launches, and soon to be gone were the dedicated Department of Defense (DOD) missions that took so much effort and expense to keep them classified. The DOD and Intelligence communities decided that going back to expendable rockets for their highly classified payloads was more expedient and cost efficient, so the Shuttle team began looking at science and engineering—and building a space station—as our goals. It is ironic that the real purpose of the Space Shuttle, as sold to NASA, Congress, and the American people when space travel was in its infancy, was to build, service, and provide transportation to and from a space station—that is what a "shuttle" is for. And eventually, that is where we came back to—first bringing cargo and crew roundtrip to the Russian *Mir*, and then building and servicing the ISS for the last ten years of the Shuttle program.

From 1988 until 1995, the Shuttle flew many payloads, including the Hubble Space Telescope (HST) and deep space probes that flew on to the planets. We built and designed a wide variety of missions, and we perfected the techniques of EVA (Extravehicular Activity) and the manipulation and capture of payloads with the RMS (Remote Manipulator System). We tried out new space structures, and we tested electrical and fluid connections that would have to be mated and demated in a vacuum (by EVA astronauts when the ISS started to go into orbit). And we flew dedicated science missions that created a growing knowledge base for researchers interested in using space for a variety of reasons.

As we began building the International Space Station and staffing it with crews that were brought up and down with the Shuttle, we made a little room in the manifest for a variety of small but important science projects. But they were all collected and put together to fly on a dedicated science mission, STS-107. This mission was slated to fly a number of different times, but it finally settled into a launch slot in January 2003. It was out of the mainstream of ISS construction and operations, and many of us were so busy with those missions that we paid little attention to it. I had been flying the Shuttle as a Flight Director for over eight years by that point, and I had quite a few missions under my belt. I was working on my next Shuttle mission, and the next one after that, and the next one after that . . . and because I was not assigned to STS-107, I took it as a chance to take a little break from the weekly routine of mission preparations and training. (Because so many people were involved in the flight, it was hard to do anything else—no one was available for meetings or training.) So I just worked in the office as needed and took a little time off. I really wasn't even aware of which day they were landing.

I had driven out to the airport on that Saturday morning to go do a little personal flying. But by the time I got most of the way there, it was apparent that the fog and winter scud were getting worse, not better. It was going to be fruitless to even give it a try. So I turned around to head home, but first I stopped at a local auto parts store to pick something up. As I walked in the door, wearing my NASA flight jacket, a customer looked up and asked, "So what happened to the Space Shuttle?" Perplexed by the question, I answered, "I don't know, what did you hear?"

His answer chilled me to the bone. "Well, they say it's falling down in pieces all over Dallas!" Just then my pager went off, followed by my cell phone ringing—and once again life was never going to be the same.

The proximate cause of the *Columbia*'s breakup was a failure of the thermal protection system—a hole in the leading edge of the wing let in superheated (3,000 degrees) gases that melted the relatively light aluminum structure. The airframe failed, tumbled, and came apart. The cause of the hole was a piece of foam that had separated from the External Tank on ascent. This foam was so light that it essentially came to a stop while the Orbiter flew through it at supersonic speed. This impact generated enough force to fracture the Reinforced Carbon-Carbon (RCC) skin of the wing near the fuselage. The technical root cause was this foam. But, once again, if you ask, "Why did the foam fail?" you eventually come back to errors in judgment by a wide variety of people throughout the agency, the program, and the operational management teams.

We were down for close to two and a half years once again. I was much higher up in the organization by then, and much more experienced in the leadership of space operations. I was therefore much more involved with fixing the people problems than I was in working on the details of technical improvements to the Shuttle system. Before I really began doing my part to fix things, I spent over two months living and commuting back and forth to East Texas, the scene of where the *Columbia* debris fell. It was a minor miracle in that the debris field was so close to home, and that it was over land. If the debris had come down over the ocean, we might never have known the details of what happened. It was another miracle that no one on the ground was hit by the nearly 80,000 pounds of debris that made it to the surface without burning up. My job was to help run the air search operations—I organized all the work of fixed-wing aircraft that were searching for debris. It was a big task, with assets ranging from powered parachutes to the NASA ER-2 (a high-altitude U-2 reconnaissance airplane in civilian colors and with unclassified sensors). It took the attention of a number of people to make sure the assets were used efficiently and safely.

All the air work, although useful, proved not to be the best method for finding and collecting debris. What worked best was "boots on the ground"—people walking gridded miles shoulder to shoulder and looking down to see what they might find. Miraculously, the Orbiter's Operational Recorder (Ops Recorder) was found on a gentle hillside nearly intact. This recorder was not designed to be a "black box" as you'd find on an airliner, but it served that purpose because the data engineers were able to extract an incredibly detailed timeline of sensors burning out and the wing failing. This allowed us to track the course of the failure. When this effort was completed, we knew why the Orbiter failed. By then we also had enough information from studying the interactions of the personnel involved in the mission to give us a good idea where the people had failed in their decision-making.

Once again, it was a matter of getting too comfortable with success and, to some extent, discarding the cries of those people on the edge of the room who had reservations about what was going on. Once again, it could be thought of in terms of Russian roulette. The signs were there: foam was coming off the tank (click), again and again (click, click)—but it never seemed to hurt us, so we figured it was safe to go again (click...). We just hadn't experienced the worst-case scenarios yet—until we did (bang!). And once the failure had occurred, there was little to nothing that could be done—the crew was essentially doomed from the start because there was no way that a rescue mission could be mounted with any sort of sane schedule or reasonable chance of success in the time available. In hindsight, many have suggested that it could have been done, but those people rarely have inside knowledge of what it *really* took to plan a Shuttle mission, and just how many chances there were for things to go wrong if you rushed through them. Realistically, we could have stood a very good chance of losing the second crew as well, simply because of something we would have missed.

My task, once I put my flight gear away and settled back into my Houston home and office, was to help out training management to make better decisions. We spent huge amounts of time training the real-time flight control teams—and even the engineering support staffs—to make good, reasoned decisions quickly and accurately during flight. And we expected our management, many of whom had come from the flight control ranks, to do the same thing. But an equally large number of program managers and NASA leadership had *not* had the benefit of growing up (professionally) in the Control Center, and we felt that since any team is only as strong as its weakest link, this time around, we needed to concentrate on training for the Mission Management Team (MMT) just as strenuously as we were training the real-time team.

The MMT was filled with good people, and everyone had a vote. But some votes were made by people who really didn't understand the topics being discussed. A representative from the Space Life Sciences Directorate, for instance—while well-meaning—didn't necessarily have the engineering background to help decide whether a limit could be changed in the hydraulic system because of a failure that was making it hard to control the temperature. They had no training in how to evaluate the presented information, and therefore no basis on which to vote other than to rely on the recommendations of other team members. The interesting thing is that sometimes this distance served a useful purpose, because it was not uncommon for those closest to the problem—the engineers and operators—to miss the big picture and forget that there were other options. It was not uncommon for a good idea to come out of left field, and the experts had to be trained to recognize it.

Our training plan for the MMT was multifaceted. It included both stand-alone training sessions (participatory classes) and stand-alone simulations. We also included the MMT in some full-up simulations for the Return to Flight missions so that they had to make decisions with the real-time flight control teams and crew who were waiting on their decision-making. Creating simulation scripts that could challenge both the operations teams and the MMT was partly a matter of looking back in the records to find cases where we thought the outcome could have gone multiple ways. We also had to look at all the changes being made to how detecting tile damage was going to be handled, and we had to make sure we trained folks in the new way of doing things. However, we always remembered that we didn't want to make the mistake of "fighting the last war." It was unlikely that our next major problem would be related to the thermal protection system, the tiles, or foam coming off the tank—so we needed to be training in problem-solving techniques, not specific cases.

One of my contributions to the training effort was a class with a twist. We wanted to make sure that the teams understood the importance of thinking outside the box—whatever box that might be. I pulled an accident case out of the aviation files and created a two-hour team-building exercise that was based on the July 2000 Concorde crash in Paris. This crash was caused by a cut tire that cascaded into a ruptured fuel tank, which failed two engines and eventually caused a crash on takeoff. Only my students didn't get told it was the Concorde accident until the end of the session. Instead, we split the students into two teams—Airline A and Airline B. They didn't know at that time that they were really British Airways (BA) and Air France (AF). Each team was given data on a tire problem that had been going on for some time. Occasional blowouts seemed to be increasing in frequency, and some had even shed rubber that dented the lower surface of the aircraft. We used real data from BA and AF but didn't mix the two.

The two teams were put at opposite sides of a large conference room and told that they had a flight leaving on this particular type of aircraft in just a couple of hours. A maintenance manager had decided he was worried enough to bring the growing tire problem up to management, and they had only a short time to decide if they were going to let passengers fly that afternoon on the aircraft, or if they would have to cancel the flight and reroute them. We also told the teams that they could use whatever resources they could think of to come up with a good solution. Without exception, every time we did this, the teams pored over the data we gave them, and they almost always decided that they didn't have enough data that indicated they should cancel the flights.

The one thing they never did was to send someone over to the other team's table to ask them what they were seeing in their own data. They were flying the same type of machine after all. If they had, they all said (in the debrief for the exercise) they would have seen that the problem was much more widespread, and they probably would have decided not to launch. Of course, this was after the "gotcha" moment when we showed them the video of the Concorde going down in a fireball. It was always a sobering moment. After we had used the training class on the program management, we did the same thing for our operations leadership and flight controllers. The lesson was that you never know as much as you think you know, and that so long as you have time to gather more data, you should do so. Eventually, you have to make a decision based on the data you have, so have as much as you can to better your odds of making a good decision.

There were nowhere near as many vehicle changes after the loss of *Columbia* as there were after *Challenger,* but there was a lot of work done on trying to once again figure out tile repair. More importantly, a great deal of effort was spent on how to inspect the Orbiter after launch. A great deal of effort was put into developing an inspection boom that could be grappled by the RMS, and then use it to put cameras underneath the Orbiter to take pictures of the entire tile surface—very much like a selfie stick. The RMS wasn't long enough, or flexible enough, to do it by itself, but the boom made the work possible. Because we generally rendezvoused with the ISS on Flight Day 3, the natural place to add an inspection day was Flight Day 2. The crew launched on Flight Day 1, then got a good night's sleep. They spent Day 2 checking out equipment for the upcoming docking and EVAs. A couple of crewmembers did the on-orbit inspection to make sure there had been no damage from the ascent.

The data was sent down to the ground via video links and was thrown to the engineering group to analyze. They spent thousands of hours each mission looking over every image they could find, comparing what they saw to prelaunch photos to make sure there were no damaged tiles, RCC panels, or loose gap fillers—pieces of felt-like material inserted between tiles that kept them from vibrating against each other. In addition to the data gathered on Flight Day 2, the crew of the ISS was trained to take a series of pictures with very long lenses as the Shuttle approached the ISS during the rendezvous of Flight Day 3. The Orbiter would approach from below the station and stop about 600 feet out. At that point, it would do a complete 360-degree backflip—the Rotational Pitch Maneuver (RPM)—and the ISS crew would take their pictures in a predetermined scan pattern using both 400mm and 800mm lenses. These images would then be downlinked and added to the pile of data already being looked at by the engineering teams. There were places that the boom couldn't reach, and the RPM filled in those blanks. As a result, we got a very good idea of what we had, and where problems might lie.

The good news is that we never had to declare an Orbiter derelict because of an inspection—but we did do at least one EVA to pluck out some loose gap fillers that might have disrupted the critical hypersonic flow pattern of the plasma during entry, causing uneven and localized heating. That really was good news, because the bad news was we never really trusted the repair techniques that were created to deal with lost tiles or missing RCC. A lot of work had been done in this area, but very few folks at the highest level got comfortable with them—at least so long as we had a rescue capability. And we always made sure that we had that capability, just in case.

It was clear that most Shuttle missions after the Return to Flight were going to be to the ISS. The only thing left on the manifest other than station assembly was a final return to HST. So for all but that final Hubble mission, our philosophy became what I called the "desert island scenario." If we got the vehicle to orbit, then we could get it to the ISS. If we discovered that we had damage that would preclude bringing it home, we'd strand the crew onboard the ISS, jettison the damaged vehicle before it ran out of cryo and prop, and then send another Orbiter to pick up the crew and bring them home.

Because we always had another mission planned to the ISS, all we really had to do was make sure that the consumables onboard the ISS would support the extra crewmembers for the time it would take to get a rescue Orbiter ready to go and up to the station. For the first few missions, we actually planned to steal the next mission's Orbiter and get it ready in an expedited fashion. After some time, and some pencil sharpening, we realized that we really could generally wait for the next mission, and fly it fairly normally, if we had to bring the previous crew home with that crew. It made for efficient mission planning and made the rescue scenario pretty routine. One of my assignments for the remainder of the program was to keep the plans for the rescue mission up to date and ready to go—something that sort of fit in with my lifelong interest and participation in rescue work. (I spent twenty-five years of my NASA career as a volunteer firefighter and officer in our local fire department—as if I didn't already have enough to do.) That made me the Rescue Flight. But since we never really got close to launching such a rescue mission, it was really just another office assignment. It was fun though, because it required a few creative touches to the planning process, and some of the usual constraints were loosened to make it happen.

The rescue mission for the final HST mission would have been much more challenging because the Hubble was in nowhere near the same orbit as the ISS. Both the inclinations and altitudes were so far off that it was physically impossible to go from one orbit to the other. I think it would have taken something on the order of all the propellant ever carried in all the Shuttle missions (combined) to get that much maneuvering capability. So in order to launch to the HST with the required rescue capability, we needed to have a complete and ready-to-go plan. Once again, as Rescue Flight, I got to build that plan. Since we didn't have the luxury of a space station in which the crew could take refuge, we had to come up with a more interesting idea.

The rescue flight for the final HST mission (STS-125) was a fully planned and paid for mission. We had a complete crew training plan for it and, in fact, the crew had to be headed to the Cape only a few days after the HST mission launched in order for such a rescue to work. STS-125 had only so many days of on-orbit life; we couldn't let the crew sit and wait for rescue on the ISS for months, because the ISS was out of their reach from the Hubble. So, as I told the program, we had to assume that the rescue crew was launching until we called them off. The mission was designated STS-400 and had its own planning process and flight preparation products. A lot of the planning could piggyback on the STS-125 planning, since it was going to the same orbit, but it still required a lot of unique mission analysis because the weights were different.

The basic plan for the rescue was that the STS-400 crew—a reduced crew of just four astronauts—would launch and rendezvous with the damaged STS-125 vehicle. That crew would have already gotten rid of the HST (hopefully fully serviced and ready to do science for as long as it lasted), and been made ready to receive the rescue vehicle. Because the docking systems of both Orbiters were not designed to extend beyond the payload bay door moldline, it was not really possible to dock them. It was a situation like how human noses tend to get in the way of a kiss. The rescue vehicle was going to have to grab the damaged vehicle with its RMS (a special grapple fixture was installed on the STS-125 vehicle to facilitate this). Once the two vehicles were grappled, the fun could really begin. The transfer of crewmembers was going to have to be done not via a docking tunnel, but using space suits and a series of EVAs.

There were seven crewmembers on the STS-125 Orbiter. Four of them were EVA specialists. They weren't a problem. The fifth Mission Specialist was the RMS operator, Megan McArthur. She had sufficient experience in EVA training that she was also not a difficult transfer, although she didn't have an EVA suit sized and ready for her. The two final crewmembers, Commander Scott Altman and Pilot Greg Johnson, had minimal EVA training because they were pilots; they were not intended to be space walkers. They would essentially be cargo in suits, carried across by other space walkers more experienced with the process. Suits would have to be shuttled back and forth, and used several times by different astronauts, sort of like the old "chicken and fox in a row boat" logic problem. It was going to take a number of EVAs, and two days, to get everyone off the damaged ship, but it could be done.

After everyone had been shuttled across (along with their launch and entry suits and escape gear), the damaged Orbiter would be cast off and sent on a reentry trajectory by ground command. The rescue Orbiter then just needed to get home. Of course, the crew size now numbered eleven souls—far more than there were seats. So we developed a special assembly of recumbent chairs and harnesses. Four would ride home on the Flight Deck, and seven would have to squeeze onto the floor of the Mid-deck. In order to be sure this would work, we tried it out on the ground with suit-qualified volunteers. It was tight but it worked. Someone looked into the side hatch while this test was going on and reported seeing "a sea of orange, all the way across the Orbiter!" It was so tight that some crewmembers couldn't reach the emergency oxygen knob on their suit, so we had to train for their neighbor to pull it for them in case it was needed.

The Orbiter used for STS-400 wasn't wasted, of course. The vehicle was already prepared to accept the next payload in line, so when it was no longer needed for potential rescue, the Orbiter was simply loaded with that payload and launched to the ISS. But STS-400 was real, and had real training; it even had a patch. We were glad we didn't need the mission, but everyone involved was a little sad to see all the work that went into it go unused. Not that we wished ST-125 ill, of course—but we would have loved to see the plan work.

When we started flying again after the loss of *Columbia*, the missions always felt different. While management was always on top of Shuttle flights before, it now seemed that the center of gravity of almost all decisions was shifted more toward the MMT. Less took place in the Control Center. That's not a complaint. The MCC personnel didn't "own" the Orbiters and the missions, we merely executed them for the program. As such, the program had the right to make decisions on the direction of a flight. But the feeling was that there was always a bit more of a "Mother May I" in the flights in the later years. Unless you had a true emergency, something not covered in the flight rules wasn't decided by the flight control team, it was put on hold until the MMT could meet. There was a sense that much of the responsibility had been taken off the flight control team's shoulders, and that the authority had gone with it.

You can delegate responsibility, provided you also delegate authority to act. You can't delegate responsibility without delegating authority. That is, you can't hold someone responsible for the outcome without giving them the means to effect change. If you give folks authority without responsibility, then you can create havoc because enabled people will be running around doing things without the burden of taking responsibility for their actions. If you delegate neither, then people are merely acting on orders without any freedom of action—an unrewarding place for the workers to be.

One of the things that changed the culture and gave more oversight to MMT was, of course, the findings that came out of the *Columbia* investigation. While the MMT was designed to be a way for all involved organizations to make their inputs into each mission after *Challenger*, it had, in fact, become a small executive committee of sorts, with a few top leaders of the program making decisions without always bringing the entire MMT into the picture. This happened slowly and without any evil intent. It simply happens when an operation becomes "routine"—you assume that since a group has made the same decision on a particular case before, it will always do so again, and so having a meeting is really just a formality. So the meetings were assumed and not held, and the folks at the top became more insulated from differing points of view to the point where some of the outlying viewpoints were never heard.

The MMT after *Columbia* was deliberately set up to avoid this concentration of problem solving and decision-making in isolation. In fact, physical signs of this effort were apparent in the development of a much larger MMT room. It was carefully thought out in terms of the shape of the table, and there was plenty of room for spectators and backup personnel from each organization who were expected to speak up and participate. Microphones were even added on the periphery of the room to facilitate this. MMT meetings were held every day, not just when needed, and the whole process became quite formalized. Formal processes are good. It is why airliners, for instance, have such a tremendous safety record these days—everyone follows the process, and the process is well thought out to prevent gaps. In experimental aviation (or spaceflight) there is always the problem that what you are doing has never been done before, so the process has to allow for improvisation and new information, and that was the challenge faced by the MMT in the last decade of the Shuttle program.

Coming up with a process that was rigid but that allowed variation was something everyone struggled with, but overall it worked out reasonably well. There were still hallway meetings and decision-making done by small groups, but everything had to see the light of day in the MMT meeting. This ensured there were opportunities for challenge from all points around the room. It worked, but it was a struggle. And the struggle was worth it as NASA's human spaceflight program moved from the fast-paced "Decide now!" culture of the Shuttle era into the "Hey, we've got plenty of time..." culture of the ISS era. The two types of program are fundamentally different because of the time frames involved—the longest Shuttle mission was two weeks, while an ISS crew was just beginning to get its bearings after a month. The process developed in the high-stress Shuttle era had to be adapted to one of a more relaxed posture that still retained the capability to react quickly when the situation required.

The flight control teams in the later Shuttle years never relinquished their responsibility to take whatever action was necessary for the safe and successful completion of the mission, and no Flight Director ever shirked their duty to execute based on their best judgment. But there was much more deference to the decisions of the MMT as the program drew to a close, and frankly there was less need to "make it up as you went along" anyway. More emphasis was put on outside team involvement when time allowed, and on waiting for things like inspection team reports to decide if things were going nominally, or if a redesign of the mission was going to be required.

It has to be observed that the ISS assembly went amazingly well, almost impossibly so for those of us who were involved in the Shuttle for its entire flight program. If you had told anyone in 2001 that thirty-plus assembly flights would all go more or less according to plan, with all the pieces fitting together as expected, and with all the EVA assembly tasks working out, we would have told you that you were nuts. But it turned out that way. Sure, there were problems, but nothing that required dumping an entire payload into space or the ocean, and there were no emergencies that turned out to be more than could be handled in flight. It was an amazing testament to the men and women of the program who grew up and learned how to handle adversity. For over thirty years they turned problems into success.

It was clear to all that after the loss of *Columbia*, the Shuttle program's days were numbered. We were directed to finish the ISS, and then move on to other goals. It was sad. In my opinion, as someone raised on exploration and pushing boundaries, it was a bad decision for the future. The Shuttle was not, and never would have been, "safe" as determined by the kind of numbers the general public expects in anything they are exposed to—auto transportation, airlines, even recreational activities. But exploration has always been about risk, and explorers are used to it.

You calculate the risk, you examine it, you minimize it where you can, mitigate it where you can't, and accept what is left. But you still go. Giving up the Shuttle was something we all worried about after *Challenger*. In fact, many of us believed that if we ever lost another Shuttle, the program would be terminated—we saw the writing on the wall that early on. America was losing its determination to do great things. When we lost *Columbia* and were told that we could still fly to finish the station, many thought that we had dodged a bullet—but we hadn't. The program was going to end and, unfortunately, the programs to come after were never well defined or properly funded. So when the Shuttle ended, the ISS was all that the agency really had left on its plate.

Commercialization of space transportation bothered many, but in the end that has been the goal all along. You use a national effort to explore, to develop, and to invent—then you take the results of those explorations and inventions and give it to the public to run with. That is how aviation worked. Here in the US, we don't fly on a national airline—we fly on for-profit, commercial airliners owned by private corporations. Space travel should become that, and I hope that it does. But it will be painful. There will be losses. And just like the Shuttle program and our two tragic accidents, the future depends not on how an organization handles success but on how it handles its failures. NASA handled its failures by applying huge resources to understand the causes, and then applying fixes to make the program better.

What we learned, however, was that it is hard to remain vigilant, to keep the processes rigid (yet flexible) enough to always force good decision-making. Success leads to relaxation, and relaxation leads to failure—and it is important to remember that you are never as good as you think you are, and that you are always susceptible to those things you haven't thought about that are waiting to bite you. If we had kept flying the Shuttle, there is little doubt in my mind that we would have lost another one, and it would have been for a technical reason we simply hadn't thought about. But if we examined ourselves, we would have noted signs that we could have seen coming. It is hard to remain focused year after year, decade after decade. It is hard—but it is required. And our only hope is that future generations of flight controllers, Flight Directors, engineers, and managers keep looking back as they leap forward—looking backward for the reminders that no matter how good you think you are, you can still stumble. The legacy of *Challenger* and *Columbia* will be that warning—and we have to hope that people will continue to heed it for generations to come.

Chapter 11

Thirty-Year Learning Curve

When you flight test a new aircraft design, the first many flights are all leaps of faith. You can analyze and calculate every possible piece of structure, you can simulate the aircraft's flight characteristics, and you can test all the systems a hundred (or a thousand) times, but modern aircraft are so complex that you can't relax until considerable time has been put on the airframe... and even then, you can (or will) be surprised by what it might do. So the fact that the Space Shuttle was declared operational after just four flights was always a cause for amusement for those of us who understood the system and what it really did. But really, we were still learning—about the spacecraft, and how to fly it—thirty-three years later on its 135th spaceflight. It was an experimental test vehicle throughout its life, and those of us who flew it knew that. Those who lived in a more political world might not have been fully aware—or willing—to accept the fact that there were always a great many unknowns that could bite you.

Of course, NASA went into the Shuttle program after successfully putting humans in space, and then sending them to the moon. They had built a space station (Skylab) and staffed it for long-duration missions three times. The men and (admittedly few) women at the Johnson Space Center had established a process by which they could send other men and women into Earth orbit and bring them home again. The Shuttle was easily an order-of-magnitude leap of complexity and aerodynamics beyond Apollo. But the techniques used to fly it were those developed during the first decade of human spaceflight, and in the days of the rocket planes flown over the high desert in the decade before that. A ground-based control center helping the airborne crew with the assistance of a stack of engineers was the way to keep eyes on all the systems and trajectories, and to help out when things didn't go as planned.

Engineering a complex system in a ground-based environment is one thing—you have lots of margin when it comes to structure and weight. When you design something to fly, such as an aircraft, you must shave off weight in order to get performance, but, generally, you can still build things pretty stout and add systems redundancy when necessary. But getting a craft into orbit (or beyond) means shaving weight down to within a gnat's eyelash of disaster in many cases. The performance difference between making it into space or not is often the difference in firing the main propulsion system for just a couple of extra seconds—a very tiny percentage of the overall burn. In the moon landing missions, the skin of the lunar module was so thin (to save weight) that if a technician dropped a screwdriver during ground preparations, it most likely would poke a hole in the cabin.

The Shuttle was built stouter than that, of course—but only enough to complete the mission for which it was designed. We were constantly working to stay within the design limits so as not to use up structural life. But it wasn't just structural life that we had to worry about. We also had to keep track of how much time was put on systems and their components. Every thruster had a life limit, every piece of machinery that rotated or moved had a number of cycles assigned to it before overhaul. And those of us in Mission Control worried about how much life was used up every time things were checked out and cycled on the ground during postflight and preflight processing. It has often been speculated that with the amount of work done on the ground, we virtually wore the vehicles out in testing—and there was some truth to that. But we certainly didn't want to forego any testing necessary to prove the systems ready for space either. In fact, we did specific testing on orbit that was designed to check things out while we were in space specifically so that they didn't need to be tested again on the ground—a lesson learned from many missions worth of processing.

It was all part of the learning process, and thirty years of flying gave us lots of time to learn. The system matured in many ways from STS-1 through STS-135, and those lessons applied to the vehicles, the crews, and the thousands of support team members all along the way. Over the decades, we streamlined vehicle processing, mission preparation, training, and the teams themselves. All of these efforts were meant to make the system more efficient without compromising safety. It was a narrow path to walk, finding where we could shave costs or make up time, and we occasionally stepped out of bounds and had to walk back. When that happened we spent more time, money, or resources until we learned a better way forward. But the Shuttle program at the end was not the same as it was in the beginning—and most of that change was for the good.

The Road to Maturity

In the early days of Shuttle flying, an MCC (Mission Control Center) team consisted of about a dozen front room flight control positions, with each of those positions fielding a backroom support team of from three to ten flight controllers. The MCC building support team, under the control of the ground controller (GC) who sat in the front room, could easily have another fifty people spread around the building to manage everything from the Mission Operations Computer to the Communication systems. Peripheral to the flight control team was management support. There were rooms for first-line managers to sit and work interfaces between the flight control teams and engineering support. In addition to managing these requests for data and information, they kept upper management informed of what was going on, and they funneled management's directions back to the folks flying the mission. The Mission Management Team (MMT) would meet once a day (or more if required) to keep all the organizations in sync with the program and deal with issues that needed management discussions, such as lengthening or shortening a mission, or doing more or less for particular payloads or customers.

In addition to all the people just mentioned, there was the Mission Evaluation Room (MER) that was owned and staffed by the Engineering Directorate. The MER staffing varied from flight to flight, and flight phase to flight phase, so it might house a few engineers looking at their systems performance or hundreds trying to solve a major problem. The MER was, for many years, situated in the bridge section of Building 30, between the office wing and the Mission Operations wing. Later, they moved to larger rooms in the Control Center itself, as back rooms shrank and space became available.

All told, the Mission Control Center was a busy, packed building with hundreds of people in place at all times, working their shift and then being replaced by a similar-sized group for the next shift. Backroom flight controllers watched over many different displays, recorded data on strip charts, kept track of the performance of all their system elements, and tracked the usage of their systems with procedures and rules. A lot of time was spent just scanning through data to make sure that everything was operating normally. Their front room operator, meanwhile, was doing less of the routine scanning so they could spend most of their time keeping up with the overall mission and coordinating with other disciplines on the way their system was interacting with everything else, as well as the mission. In general, we used to say that the back rooms were there to serve their system, and the front room was there to serve the mission.

One of the goals of the Shuttle program was to continually make space more accessible to humankind, and to do so at a reasonable cost and use of resources. This meant that throughout the three decades of operations, we were constantly looking for ways to save on overall effort. One of the most obvious ways for this to happen in the Mission Operations and MCC world was to cut down on the number of people supporting each shift and each flight. Early on, just about everyone in Mission Operations had some assigned role in every mission. This was done partly to have as many eyes and brains engaged, and partly for morale purposes. It was truly lonely in the office when everyone was working a mission and you were left out!

But as time and missions wore on, the grind of constantly flying with everyone involved got old, so flight controllers themselves started looking for more efficient ways to do things. The first was that "hangers on" were generally dropped, because with an increasing flight rate these folks needed to be back in the office getting ready for the next mission. There were still legitimate spots for those doing on-the-job training, but we no longer had assistants for the backup person for any system.

It was easy to reduce the count of people who were just there to watch, but we also needed to try and reduce the number of folks who were simply watching numbers change. This was done with the development of smarter monitoring software—a precursor to artificial intelligence (AI), which was not really that smart back then. Whereas the early MCC software basically decoded, calibrated, and displayed raw data to the flight controllers, later developments actually provided smart monitoring, alerting flight controllers to trends that were unusual for the systems. This freed controllers from a lot of boring and routine monitoring, which in turn allowed them to work on procedure and timeline updates—things that required higher level intelligence and innovation. The control teams began shrinking when these monitoring aids became available. By the end of the program less than half the flight controllers were present at any point in time in the Control Center than there had been at the beginning.

It's worth noting that the ISS program eventually reached team-size reductions never achieved in the Shuttle days. This was done by actually combining front room positions at night or during slow mission times. The Shuttle program would occasionally release a front room position that had nothing to do on orbit, or when their systems (like EVA [Extravehicular Activity] or RMS [Remote Manipulator System]) had been stowed for the mission. But the Shuttle program wasn't like ISS on a slow Saturday night, when only a Flight Director, ground controller, and one or two super systems controllers were on duty.

As flight control teams began to shrink, so too did their demand on facilities. Back rooms were consolidated until we basically had a single large back room for the systems folks, another for the flight dynamics people, and one more for the flight activities folks. Bringing these back rooms together added to the synergy between disciplines, because they didn't have to talk to someone on the other end of a communications loop. Instead they could just lean over the console to the other controller, draw pictures, and discuss things face to face. These consolidations continued as the Shuttle team moved to the new Control Center (really an extension to the old building), and the entire architecture of the hardware changed. Gone were the P-Tubes, and we welcomed the ability to use office automation tools (PCs and the internet) to communicate. Documents that used to have to be duplicated and walked around by messengers now were available to everyone instantly and we no longer needed to have huge recycling bins to handle all the waste paper.

A good example of how automation helped was how we reviewed messages going up to the crew. In the early days, this was done via an actual TV camera in a back room. The message was typed up and put on a flat bed, and the camera was suspended vertically above it. The team would be called to a particular TV channel, which they'd bring up on their console. Then a person in the back room would grab a pen and mark up the document per conversations on the communications loop. We called this system Mister Hand because the person with the hand was simply marking up what they heard as it was said. Sometimes Mister Hand would mark up something, then another person would correct it, and then a third would come up with something altogether different... until the scribbles got so hard to read that the document would disappear, a clean copy would appear in its place, and we could start over. I don't recall any obscene gestures being shown by Mister Hand to the team due to frustration, but I wouldn't be surprised if this happened.

Later on, of course, we could simply have everyone look at the current version of the document on their PCs, the messages having been placed in a folder for everyone to see. You'd submit your markups digitally, the corrections would be made by the appropriate people, then the document would be redisplayed. If there were conflicting inputs (which were rare), then they would be reconciled on the Flight Director loop by the team.

The new technology cut back considerably on the support teams for the Flight Activities Officer (FAO) and helped streamline operations when the program had support teams around the globe. By the third decade of the Shuttle program, the teams were down to a reasonably small size, productivity was up due to automation, and more people were simply placed on call in case they were needed during a flight. Missions were no longer a case of all hands on deck, and creative engineers were busy working on future missions and spaceflight plans.

If It's Exciting, You've Done Something Wrong

It's ironic that while spaceflight is supposed to be exciting because it is exploration, doing brave new things, and advancing humanity beyond our planet, those of us who actually do it try to keep it as *unexciting* as possible. Now we say that with a little wink, a touch of tongue in cheek, but that's just because there are different meanings to the word *exciting*. If spaceflight itself is exciting because of what it is and where you're going, that's great. But if it's exciting because things are going wrong, or happening too fast, or you don't know exactly what's happening... that's a bad thing. Mission Control has an unofficial motto: "Making the most exciting thing in the world as boring as possible for over fifty years!" We like high achievement, but we want everything to go as planned—and many people think that is boring. And that's okay.

We like things to go as planned, we want to stay on the timeline, and even though we train for excitement, improvisation, and talented "genius" troubleshooting, things go much better on the real day if we don't have to do any of that. You don't want fires, leaks, or control failures. You want to make it look so easy that everyone else wonders why you need to spend your life training to make it happen.

Making things boring when you are ready for anything is a matter of keeping your margins high. It's sort of like training to walk the tightrope but having a net below you so that if you do make a mistake, it's not fatal. In fact, we like to make things so boring that sometimes it feels like the wire is just sitting on the ground and we're just walking along it. There's still the chance that you will trip, but at worst you'll skin your knee.

Why is boring good? Why do we strive to make things as sedate and tranquil as possible? Well, our goal isn't to go to work and have an adrenaline rush caused by the unknown. We aren't daredevils, looking for a thrill. The thrill comes from executing the mission—getting done exactly what we set out to do. If what we are doing (the mission) is exciting, rewarding, and future-altering, well that is where the excitement comes from. Making a critical call that makes the mission succeed is like winning the Super Bowl for a flight controller. It's all about making history—even when your name never appears in public. Knowing that you were part of it is satisfying enough.

I look around my office and see mementos of missions. The Contingency Control Panel for the Spacelab Instrument Pointing System—the training folks gave that to me. I have a little Lucite block with a tiny piece of Shuttle tile that came off the *Columbia* after STS-1—a memento from being a co-op during that momentous time. There are crew plaques from all the flights that I led—the names of all those astronauts bring back the memories of faces and the voices of folks who did good work in space for the betterment of humankind. All these things, and the many photographs, remind us that our teams did good—well enough to make history happen.

Flying the Shuttle was full of excitement in and of itself. We didn't need to make it any more exciting by having things break. The goal was always to make the mission successful, not to come home with our hair on fire.

If You're Not Ahead, You're Behind!

Time is the single nonrenewable resource that we have. You can buy more spaceships, you can hire more people, you can get more funding—but once a moment is gone, you can never get it back. When you are hurtling along at 5 miles every second, time passes very quickly, as do opportunities to make things happen. You can't ask for a rewind; you don't necessarily get a chance for a do-over. You must be ready when the time comes, and you must make sure that you are ready to be right. And since most things that require precise timing require you to be right on time, the only way to make them happen is to be ready in advance. Hence, what I have always told my teams: if you're not early, then you are already late.

Spaceflight is horribly complicated, horribly expensive, and horribly unforgiving. It involves hundreds of thousands of people, and when you think about it, you often have only one chance to do something correctly in a mission. If you're the one person who misses their cue, you could waste all that money, time, and effort—simply flush it down the drain. If you're the kind of person who freezes up when faced with that reality, then being involved in such a position probably isn't for you.

When we flew the SRTM mission to map the earth, the complete map depended on getting three passes over every bit of dirt that fell underneath the orbit of the vehicle in the remarkably short time of ten days. One hundred fifty-six orbits—that was how the geometry of the swath, the orbit, and the altitude worked out. When we came "feet dry" (crossed a coastline from water to land), we had to be mapping. The Orbiter had to be in attitude, with the attitude rates low enough that nothing would disturb the radar beam. At that same time, the payload had to be ready to start taking data—tapes had to be in the recorder, the recorder had to be activated, and all the many and various commands to activate the radar had to be sent. That was a lot of things that could trip you up. If any one of them didn't get done on time, then you missed the pass and the map would be less than complete.

Because timing was often so critical, the only way to make sure that you were going to be ready (for anything) was to complete all the steps that could be done early. You'd front-load activities so that all you had to do was send one last command, or flick one last switch, when the time arrived. This was the way orbit maneuver burns were conducted—everything was set and ready to fire well before you got to the last ten seconds before the burn. At five seconds prior to the time of ignition, the computer would prompt the crew one last time with a message that didn't actually say, but meant, "Are you sure?" The crewmember would then hit a single key to do the final arming of the burn software, and the computer would fire the engines on time. The full burn prep procedures could take five minutes to set up, but you did them all well in advance, just to make sure.

The same thing was true in mission planning. I expected people to front-load their work and get things done before the actual deadlines. This was often because the goalposts frequently moved—payloads made changes, the program made changes, the timing of the mission changed, the rendezvous target had changes—and all of them came after you had done your work. So that meant you had to do it over. The only way to deal with these late changes was to be early with your normal work so that you had extra capacity to work the new stuff when it came in.

I worked a huge number of planning shifts in my days as a Shuttle Flight Director. I sort of liked the predictability of the process, even though the actual work we'd do could be very unpredictable. The one thing that all those shifts had in common—when I was leading them—was that I refused to ever be late. I expected my team to have their inputs done early, have their reviews done early, and be ready for uplink of the new daily plan early. That way, there was no final rush, no panic, and no missed deadlines. Be early and you won't be late.

The First Answer Is Always Wrong

Because spaceflight is fast-paced, and because we always wanted to show everyone that we were on top of what was happening and had quick reactions (and minds), there was a tendency for flight controllers to want to solve every problem within milliseconds of seeing something happening in their system. This tendency toward fast answers is admirable as such—but it also leads to problems, because in most cases of real failures the first answer is always wrong. If you take a little time, look around, see what else is happening, and put what you see in context, you almost always find a better answer or better path to follow than you initially thought. Flight controllers got caught in this trap over and over again, which is why we tried to encourage fast thinking but slow action.

Part of the problem with making snap judgments was the way that we trained. In part-task systems trainers or full-up integrated simulations, the instructor put in a failure and the student was expected to recognize it and react to it. There was generally a single answer to each problem, so the only measures of "goodness" were if you got the right answer and how quickly you got there. Hence, we "trained in" the trait of being fast, of having the right answer as quickly as possible. There is nothing wrong with this per se, but it isn't how the real world of spaceflight works.

Generally speaking, failures seen in flight are fuzzy—we see some symptoms that we have never seen before, and nothing bad happens right away. A pressure starts fluctuating, a temperature begins a funny cyclic behavior, or some digital bits start showing weird patterns. We look in our bag of failures that we had envisioned before the mission, and that we had experienced in the simulator, and we try to match what we see to what we trained for. Our brain then comes up with the closest match, spitting out an answer. The problem is that we are generally trying to fit a square peg in a round hole, because what we trained for is not what is actually happening. The strange pressure fluctuation isn't a leak across a pair of seals in an accumulator but is simply the variation of pressure due to temperature changes induced by a nearby cooling line that is alternately freezing and thawing because a payload box mounted on the cooling system is operating intermittently. The temperature is cycling in a way we have never seen before, due to shading by a solar array on the Hubble Space Telescope temporarily mounted in the payload bay for repairs—and our attitude changes weren't taking that into account at the altitude we are flying. And the funny digital bit pattern is there because someone programmed the wrong decoding table into the MCC workstation, so the data is gobbledygook—nothing is wrong on board.

There are many ways in which you can be fooled into thinking that you know what you're seeing. But if you come up with a quick answer, and if you spit it right out to the Flight Director (or the Flight Director tells the MMT what is happening too early), you will almost certainly be wrong. Being the first with the wrong answer is never a career-builder—and it doesn't help the program, the mission, or the flight either.

Now of course there are cases where you must act *now*, and you have to be *right*. That is simply the game in a high-risk, high-speed flight environment. If you have to take a course of corrective action immediately, then you want to try and take the most conservative action you can that will leave you options for recovery if (when) you come up with the *real* correct action. Head in the direction of "goodness" but be ready to change course when you gather a little more information. This is an awful lot like saying "take your best guess." But sometimes the best *first* answer is an informed guess.

The important thing to remember is not to be paralyzed by the fear that your answer will be wrong but to understand the tendency toward rushing an answer when we *think* we know what is going on—and to fight the tendency to be the first with a wrong answer. Know this is going to happen and temper your response, tailoring your quickness to the time you have available. Know when you need to act, and then take just the action needed—and always leave yourself an out so that you can take the correct action when you figure out what that is.

Never Make the Right Decision Too Soon

The sister axiom to "The first answer is always wrong" is to never make the right decision too soon. All problems have a time when their effect becomes critical. Unless it is something actually exploding, you almost always have a little time to figure out what is happening, what the effects will be on the vehicle (and the mission), and what the potential actions are that can fix (or mitigate) the problem. So the first thing you have to determine in the first milliseconds that you see a problem is "How much time do I have to solve this?"

Being the first one to get to the wrong answer never gets you any points.

Even if you look up and see that you are about to drive over a cliff, you probably have at least a little bit of time to figure out if it would be better to go right or left before you turn the wheel. Remember that old pilot saying that goes, "In an emergency, the first thing to do is to wind your watch"? The adage reminds us that we should never just react. In most cases, if you're not already dead, you should be deliberate, figure out what is happening, gather some data and information, formulate choices, and only *then* decide. "Winding your watch" is a metaphor for taking this pause.

Many folks see this piece of advice and ask, "Why not just choose an answer and get on with it? What's wrong with being initially decisive?" Well, the bottom line is that we never have enough information to make any decision. By "enough" I mean all the relevant data that goes into choosing the best answer in any given situation. We can always collect more data, always ask for more clarification—right up until the time that the decision *has* to be made. Making good decisions is about knowing as much as you can about the alternatives, the desired outcome, and the place you find yourself in. Without building a good knowledge base before making a decision, what you are really doing is just guessing. A guess is really not a decision—it's just a stab in the dark.

By agreement with the Shuttle program, the flight control team (through the Flight Director) had the authority to take whatever action they felt was necessary for the safe and successful completion of the mission. This agreement was basically the flight rules document. If anything happened that is outside of what was covered in the preapproved rules, then one of two things occurred. If there was no time to consult anyone outside the flight control team, then the Flight Director made a good decision based on all the information they had at the time. But, if a quick response was not required, then the decision got shuffled off to the MMT. The team had to put together all the necessary data, options, and recommendations so they could brief the MMT so that the program management could decide what to do.

The key thing to remember in either of these cases is that we should always take as much time as we have available to make the decision—especially if the decision is irrevocable—so that we don't head down a dead-end path. What do we wait for? More information, new ideas, or a change in the situation. Any of those could affect what we eventually decide to do...so why make the right decision too early?

There Are Always Alternatives (Maybe Not Good Ones)

I always taught flight controllers that before they made any decision regarding a course of action, they should come up with at least one alternative. Sometimes, that alternative might be to do nothing; other times, it might be to do the opposite of what their first inclination might be. In any case, even if you are facing a problem that requires a fast response, taking just a moment to think of at least one alternative might just make you realize that you were headed down a poor path. Or, as I also told them, the alternative might be so stupid as to defy all logic—but that just proved your original idea probably wasn't that bad, and you can proceed with confidence.

If there are no alternatives to a decision that you are making, then truly it isn't a decision at all, is it? You are simply executing the next step of an inevitable process. Stopping yourself to come up with an alternative will make you feel better either because you now realize that you have made a choice between multiple options or because you realize that the decision was out of your hands anyway.

It was not uncommon in the Shuttle world for a discipline to have a problem, and while they were describing it to the Flight Director, another discipline would pipe up with an alternative because they saw other data that made it clear there were larger issues afoot. The MMACS (Mechanical Maintenance Arm and Crew Systems) Officer might have lost a heater on a hydraulic line, for instance, and it turned out that this tipped off the EGIL (Electrical Generation and Illumination) controller that a power bus had failed, and that this was, in turn, going to affect other heaters that belonged to other operators. This type of discussion was important for the flight control team, because by exploring options they determined the real scope of the problem.

The day that the *Columbia* broke up on entry, the first indication of the problem came from the MMACS console when he lost a tire pressure indication on the left landing gear. Usually, an indication like low tire pressure would be interpreted as an instrumentation failure: because there were multiple measurements for each tire, if one indicator went away suddenly, it was more likely that the instrument failed than the tire leaked. On that day, there were multiple instruments showing a loss of pressure. But the tire hadn't actually leaked; the breach in the wing had burned through all of the wiring. It was quickly determined that more instrumentation was failing and, in this case, there were no alternatives. But the discussion that went on in that short period of time was exactly the kind of thing that made one think of alternatives—if it wasn't the tire going flat, then what was it?

The bottom line was that with a vehicle and operation as complex as the Shuttle, if you couldn't think of at least one alternative to any course of action, then you probably weren't creative enough to be working on the flight control team anyway.

Trust Your Data—But Never Believe It

Of course, all this talk of looking for alternatives, taking more time to look for better data, and not reacting too quickly depends on data. We have to trust our data, but we also need to be suspicious of our data. It's a paradox, one that you lived with in the Shuttle program every day. You always had to ask yourself, "Could the data be lying to me?"

Data from the vehicle was obtained through electronic means. A temperature measurement, for example, started as an analog voltage that would be converted to a digital word that computers could use, and that word would be inserted in a stream of data words that then were transmitted to the ground. The ground computers would then decode the data stream, breaking it into its component word. The different console display applications (programs) would then calibrate the word and display it. In this case, it would appear as a temperature to the controller. It was a long route for the data to take, and all sorts of things could happen to it along the way.

Maybe the temperature sensor itself was experiencing a mechanical failure. Perhaps it became unstuck from whatever it was supposed to be measuring. If so, then it could have been subject to some corrosion that could have changed its electrical characteristics such as its resistance. Or perhaps the voltage supplied to the sensor might have varied, altering the accuracy of the output. Or maybe the black box that converted the voltage to a digital word was having a problem. All along the way, there could have been communications failures that froze the data on its last value. Whatever the failure, the flight controller looking at the screen might have seen a steady temperature, when in fact it was dropping or climbing.

So the first rule of trusting your data was to never trust your data... at least not one single data point. The trick was in understanding how instrumentation could lie to you. A temperature measurement that went from 100 degrees to 0 degrees in an instant was much more likely to be a failed instrument than an actual temperature drop. If you watched it drop steadily over time, then it was more likely to be an actual temperature drop—most of the time. In order to make sure that critical data was being monitored, the normal design would include redundant temperature measurements—a second transducer right next to the first, with its own data channel. But there is an old saying that goes, "A man with two watches never knows what time it is." In other words, two measurements could help, but you couldn't always know which was correct unless one failed high or low.

A lot of the time we were flying with new science or engineering payloads that had never been tried before. As a result no one really knew what the data might look like before we got into space. In these cases, of course, we would generally be suspicious of everything and tend not to make snap decisions based on anything we saw until we gained some experience with it. This was fine for things that didn't affect vehicle safety. But if vehicle safety was at issue, then we relied on our flight controllers to be experts in their field. They had to draw from a wide range of experiences and apply their personal knowledge about how their system might work. It was why we had engineers rather than technicians watching over the vehicle—with a new and complex machine, you never knew exactly what you were going to encounter.

Overall, we found that the data was pretty reliable. But we always needed to take a moment when we saw something we didn't expect to ask how the data might be lying.

In God We Trust—All Others Must Bring Data

Even though we were always asking ourselves how the data was fooling us, the fact of the matter was that data was king. In any argument or problem-solving session, the person with good, hard, solid data was going to win over someone who was just speculating. The Mission Evaluation Room (which housed the engineering support teams for the missions) had a sign over the door that said "In God We Trust—All Others Must Bring Data!" We didn't decide things based on guesswork.

The Shuttle was a multi-billion-dollar program responsible for priceless national assets—and the lives of their crews. Guessing just didn't cut it. It was okay to run up against a problem that no one had ever seen before, or that had no apparent solution. In such cases we did what we were paid to do—find the reason behind what was happening and come up with potential solutions. It was hard work, sometimes costly—and it could take a long time.

A good example of this was the "summer of hydrogen leaks." It began with a launch scrub due to a high concentration of hydrogen in the aft compartment of *Columbia* on the launch pad. The launch attempt was for the STS-35 mission carrying Astro-1 in May 1990. The fear (totally rational, backed up by testing) was that if you lit off the engines in that environment, you could have quite an explosion. The Shuttle was drained of fuel and the crew was sent home. The engineers scratched their heads looking for the source of the leak. It was not obvious—there were no smoking guns that showed where the leak had occurred. The first response was to tighten a few things and try again. New countdown—same result. And so began the long summer of troubleshooting that involved two Orbiters (*Atlantis* was tried after *Columbia* was rolled back to the hangar), numerous disassembles, destackings, reassemblies, and restackings. There were tests, tests, and more tests. Thousands of engineers got involved, and massive fault trees were worked through.

The problem was that they were dealing with multiple failure modes from a few different sources. The hydrogen leak was different on each vehicle. The sources were eventually found and repaired, with numerous process changes as a result, and the Shuttles returned to flying for a long time. But all the work was based on testing and analysis—not guessing, not wishing, and not hoping for a better result. The work was data driven. I remember sitting through countless meetings, with reams of data being presented, trying to sort out all the different variables. It took a large tiger team to eventually sort out the issues, but that is what engineering is all about.

We expected no less of our operation engineers. While we had to make decisions at a fast pace while the Shuttle was in flight, we expected decisions to be based on data—not guesses. That meant flight controllers had to study the history of their systems and remember all the various tests and results that occurred over the decades of Shuttle design, construction, and testing. There was no substitute for knowing your system intimately—it was the price you paid for walking through the Flight Control Room door.

Better Is the Enemy of Good

We always strived for perfection in anything we did, but we were always realistic about the fact that perfection really isn't attainable. Eventually you had to figure out in Flight Operations that "good" is usually sufficient and that "good enough" was the minimum goal to shoot for. Those with the luxury of infinite time can keep working on something until it is perfect, but if you are trying to achieve a goal by a certain date, then you have to understand when you have reached the point where you can safely get there without further time and effort.

The trick to getting a job done with excellence is knowing when you have reached a level, a bar, that satisfies all the requirements that you set out to achieve. If your goal is perfection, then you are not going to get much done—you will forever be striving for your first success. If, however, you know where your bar is—and it is okay to set it high—and you reach that point, then it is time to declare success and move on to your next challenge.

We faced this fact daily in MCC. While many people—in both the operations and engineering community—were searching for perfection, those of us who were tasked with carrying out a mission had to be realistic and move on when we had reached our bar. A simple, but useful, example was when we got the Execute Package ready for the next day's activities. It was scheduled to be uplinked just before the crew woke up. If it wasn't ready, then either the crew would be doing work based on the previous day's plan (which could be wrong and lead to wasted effort), or they would be stuck twiddling their thumbs until we got the plan up to them. Neither outcome was conducive to the best use of precious on-orbit time. Therefore, the goal with the Execute Package was to get the best product up to them in the time we had available, and to make sure we didn't send the crew anything that was just plain wrong.

There were times when we simply didn't have all the work done, or all the answers that we needed. In that case, we'd put up what we did have and a skeleton for the rest. You needed to know what the crew needed to get started and what you could keep working on. The crew needed the morning activities in detail, and they needed to know what they could expect for the afternoon. And so long as they got the details by the time they needed them, everything would work out okay.

Trying to make things better than they needed was appreciated— so long as it didn't slow down the process to where nothing happened and the crew didn't get what they needed before they needed it. Yes, you could always make things better, but if that happened at the expense of getting things done, then it was worse—not better.

The Trick Is to Remain Flexible Without Becoming Completely Limp

It takes a lot more than just technical ability to accomplish a difficult task (such as flying a spacecraft) in the context of a very large organization—it takes effective leaders who have experience with managing people. If all you had to do was keep the equipment working, then the work would frankly be easy. The headaches and pain come because we have to serve many different interest groups with different demands and expectations. Some of those interest groups might want to achieve the goal in a different way. The art of leadership in this environment is in keeping as many people satisfied as possible, while keeping everyone inspired to reach the stated goals of the mission. One key to doing this is to remain flexible. Becoming hard-headed on mission requirements when you don't actually own those requirements is a sure road to unhappiness.

The design phase of any mission is where flexibility is especially important. Mission goals are often fluid, launch dates are approximate, timelines are vague, and the options for trajectory choices are many. If you set your heart on one option too early, you'll find yourself swimming against the current when new information—oftentimes from international partners—comes to light. Yet remaining totally flexible doesn't show a lot of leadership—you have to address changes while keeping a reign on what is reasonable to change and what is not. The trick, as some sage opined in the Flight Director Office (I have forgotten exactly who), was to "remain flexible without becoming totally limp."

Becoming totally limp might have been the way to keep the program (Shuttle, ISS, or something in the future) happy—they owned the requirements, and therefore they were the ones who directed what needed to be done. But it was the Flight Director's job to implement those requirements by engaging his or her entire army of operations people—designers, time-liners, trainers, and flight controllers. And there was no better way to tick off the troops than to issue marching orders for one option on one day, and then send everyone off on a different path the next—followed by pointing them in an entirely new direction the day after that.

A good example were the timeline decisions in the ISS era. Launch days were very hard to predict more than a couple months in advance. Sure, you had a planned launch date—everyone needed a target to work toward. But I had a rule of thumb that outside of three months, the launch date would slip a week for every month between "now" and launch; and inside three months, it would slip a day for every week remaining. That seemed to work out pretty well. It acknowledged that we all had to be flexible when it came to the launch date. (The only way to absolutely guarantee that the launch date wouldn't slip was for the Lead Flight Director to plan an expensive, nonrefundable vacation for the weeks after the flight well in advance. Invariably, the flight would slip right into that time frame and never move again.)

Early on in the ISS program, we put a great deal of effort in the detailed design of the timeline for a particular launch day. Then, as we got closer to the flight, the day would slip. The FAO world would burn the midnight oil building an entirely new timeline. We'd settle on that new timeline just as the launch slipped again—but not enough to reuse the first timeline. So a third would have to be built. Then a fourth, and a fifth, all with folks working around the clock. The system was fairly rigid, and so the replanning was expensive. But we learned from this, and eventually we created a planning process that was much more flexible. We kept the hard and fast rules, of course, but the FAO planned timelines based on different trajectory events and launch dates right from the start. It was being flexible without becoming completely limp. In fact, it became so flexible that generally they didn't even build a detailed timeline until after we launched. The FAOs by then had so many canned timelines in their computer that it was a simple matter of pulling the right template out of the bag and letting the computers fill in the details.

Flexibility in one's thinking was also important—it was much harder to change a mind that had been set than one that was open to change. And change always came—changing requirements in flight, changing priorities, changing rules. You didn't need to be a fan of change, but you did need to learn to accept it, and to run with it. That can be hard to do when you live in a rule-based world, but it was something you came to grips with—or you found another job. I learned this gradually, but I finally reached the point where I told my controllers to relax when it came to the requirements. We didn't own them—we implemented them. We had flown enough missions that we had templates for just about anything that the program might ask us to do. And as I always told them, "Unless it is illegal, immoral, or fattening, we are there to serve the program—so let's do the best we can to give them what they need!"

Chapter 12

Leadership in a High-Risk Business

It has been almost four decades since I first walked through the doors of Mission Control. Across those many years, we flew the Space Shuttle 135 times, began routine space station operations with the Russians, and constructed the International Space Station. Along the way we deployed countless major science satellites and conducted thousands of experiments on dedicated science missions. Despite the wide variety of missions flown during those years, some things remained the same—the dedication and commitment of the men and women of Mission Control. It is rare to find a group as focused on doing a job well as the people that plan, train for, and conduct human space missions. While their spirit and commitment remained a constant, there can be no argument that the techniques and organization of the work done by these many thousands of professionals changed over time.

So it is useful to reflect on the nature of many of those changes, muse on their effects—both good and bad—and, hopefully, capture many of the lessons taught and learned by the multiple generations that have called Mission Control home. There are many different opinions about what constitutes "good" flight control, or even a "good" Flight Director. But most agree on a few key principles—not the least of which were the core values inherited from the early days of spaceflight. These principles were kept foremost in the minds of every generation of flight controllers as they passed through the various vehicles and programs that were supported by the Mission Operations Directorate (MOD). In the end, it made no difference what generation you were or which vehicle you flew—you held true to the core values because you were part of a unique team that did great things.

Core Values

Much has been written on the core values of Mission Control and its people over the years. But if I had to choose one essential fact that sets Flight Control Room (FCR) flight controllers apart from everyone else—no matter what the venue—it is that they were always expected to be the smartest person in the room in their area of expertise. Whether it was a systems controller discussing a failure mode or operational procedure, a time-liner building a flight plan with program and payload elements, or a Flight Dynamics Officer helping to decide on a trajectory, the Mission Operations' traditional goal was that no one would know more than the operations team. History has shown time and time again that flight controllers have had to lead other organizational elements to an answer that MOD had already figured out. Of course, we all know that there are always people smarter than we are—which is why all the NASA organizations are involved in missions. But the goal is to drive every one of our flight controllers to a high level of excellence.

FCR flight controllers grew up as engineers who understood the systems because they were immersed in the development of those systems. In the early programs (Mercury, Gemini, Apollo), the systems were developed from scratch on a timeline that meant that everyone was part of the team, from conceptual design through the last flight of the vehicle. Flight controllers helped design the redundancy, they defined the data and command systems, they produced the procedures and flight rules—all from scratch. The very concept of the system-level drawings—produced from blank sheets of paper by the flight controllers themselves—came from the need to interpret schematics to show *how* a system worked, not just "how to work it."

Flight controllers learned the details of system operations better than their engineering counterparts because they not only developed the operational documentation and followed its development and production, but also because they trained constantly on failures and failure modes. While many of the training failures produced by simulators were of dubious quality, the very fact that flight controllers had to ask themselves, "How can it be doing that?" forced them to understand the systems at the bit and bolt level. No good flight controller left a stone unturned when evaluating a failure signature. They thought outside the box before the concept was popular. More importantly, they developed the habit of picking a reasonable course of action and seeing it through, even before the detailed failure analysis had been performed. This was because they *knew* that they were expected to be the smartest person in the room, and they also knew that they were the last backstop against catastrophic failure when the spacecraft was hurtling around the planet.

Flight controllers were never trained to have this trait—they trained themselves because they saw the trait in their mentors and progenitors. They understood that they were expected to "know it all," and their superiors made it clear that nothing short of a mastery of their subject was acceptable. While this worked well with short-term programs such as Apollo, it became more and more difficult with a long-duration program such as the Space Shuttle or the International Space Station. Quite frankly, the young flight controllers who came along twenty years into an established program had no way to simulate the learning that the early folks experienced—the drawings, procedures, and flight rules were already written. They were forced, instead, to learn the system from books and lessons, and none of that required them to use their creative talents to develop a deep, gut-level understanding of the material. They learned "how to work it" rather than "how it works." Slowly, over time, the leadership and management of MOD began to accept this as "good enough" such that it became the new standard by default rather than by design.

NASA and Leadership

Leadership is vital to any large human endeavor—it is the motivating factor that brings large groups of people together to accomplish great things. Countless books have been written on the subject, but when you boil the topic down to its bare essentials, these are the common traits of successful leaders:

1. They have a vision (mission)
2. They are focused on accomplishing that vision (mission)
3. They inspire others to believe in that vision (mission)
4. They enable their followers to accomplish that vision (mission)

Without a vision, you have no leader, without followers who believe in that vision, you have no leader; therefore, the leader must be focused on serving his followers in order to enable them to get the job done. A good leader looks forward to see the vision accomplished, and he looks back to make sure his people have what they need...but he doesn't care about looking *up* to see if his superiors are happy. His superiors should be happy when he accomplishes his mission—or he is working for the wrong superiors.

This will no doubt be controversial and ruffle more than a few feathers, but it is my expert and studied belief that NASA expended too much effort breeding out leadership qualities and breeding in management techniques—and this is one significant reason for the declining fortunes of the agency over the years. Leadership and management are two different things. Leadership, as stated above, is about inspiration—inspiring a vision in others. Ideally, a person in a position of leadership would also be a good manager—but truthfully, a good leader can have others manage for him if he is inspirational enough. A good manager might be able to put resources in the correct places, but unless he can inspire, all the resources in the world will be for naught.

The great leaders throughout history have always been inspiring. A good leader needs to inspire their followers to share the vision, and then get out of the way so that they can accomplish it. You can't do great things without a vision! Good managers need to either partner with good leaders or be visionary and inspirational themselves or nothing bold can happen. People know a good vision when they see one, and they know when they are not being inspired as well.

In the "old days" of NASA, the agency was filled with leaders who were able to inspire the workers to do great things. They were willing to take responsibility for decisions, large and small, to keep the ball rolling and momentum high. In the later years of the Shuttle program, NASA's management techniques tended more toward deferment of important decisions, and the workers sensed the fear of taking the bold steps that are inherent in the process of spaceflight. Good leaders of the past took the attitude that they might be fired later, but they knew the right course that needed to be followed to achieve their vision and they were willing to take personal risks to achieve them.

Good leaders know that they are nothing without their team, and so they spend most of their time looking at the team members to see what they need. They are constantly engaged with serving their team so that the vision can be fulfilled. They are aware that their team members need their support, and these leaders are able to see that this is the way to accomplish their goals. Managers, on the other hand, tend to be oriented to pleasing their superiors—they are constantly looking at ways to make their boss happy, and this frequently happens at the expense of the workers. Managers are oftentimes (not always, but more than not) working to succeed at climbing the ladder rather than achieving a vision, because much of the time they don't have a vision.

Trust is vital for leadership to succeed. The workers need to trust the leaders—and they can only do this if the leaders trust them. The quickest way for leaders to lose their followers is to not trust them, to micromanage the team and prevent the team members from using their originality and creativity freely and with recognition. Good leaders do not worry about their workers talking to the leaders' superiors directly. In fact, they take pride that their team members take the initiative to do so. Managers, on the other hand, feel the need to control and oftentimes throttle even the minutest bits of information through themselves, making sure that they are a single point of contact to upper levels of management. This mistrust of the workers is obvious and a huge detriment to morale. Oftentimes sold as a way to "show a common voice," it rarely is. It reveals, instead, a lack of leadership training and skill. It was pervasive within NASA's management core in the 2010s.

MOD was not immune to this decline into "control by management." In the past, MOD was a significant source of leaders for NASA. Later, it was a significant source of managers. NASA's inability to bring programs to fruition was potentially attributable to this trend. Certainly, the political climate is a direct hindrance to the success of NASA's objectives, but this should not be used as an excuse. If you study the history of Mercury, Gemini, Apollo, Skylab, Shuttle . . . they *all* had significant political opponents and teetered on the brink on numerous occasions. They were pushed through and saved by strong leadership—not by management explanations, bullet charts, and telling politicians what they wanted to hear.

An organization that wants to be successful needs to continue to encourage people to study successful leaders—and to identify good leaders for those who are trying to pick their role models.

Mission vs. Organization

Like most large organizations that started out small, NASA (and its sub-organizations) became a self-eating watermelon. In many cases, NASA became more concerned with organizational survival than with mission execution. It can be hypothesized that this was part of the conversion to a "business" outlook, in which the survival of the business is assumed to be a top priority. But in the case of a mission-oriented organization, such as NASA was when it was formed, the goal is not organizational survival but mission execution. This is one of the reasons that we became so conflicted in program management—we did not have a clear mission and, therefore, we simply tried to make the organization survive until the next mission came along. But if the organization isn't right for the mission, then the mission suffers as the organization tries to refit the mission.

I would submit that organizational survival should *not* be a significant driver in a group that is trying to execute bold, important missions. In fact, I like to tell people that I really had no emotional attachment to, or interest in, MOD—the Mission Operations Directorate. This shocks many, and is, in a way, a bit of baiting (because I do care greatly for the people and the traditions of MOD). But I didn't come to work in Houston to be part of MOD—MOD didn't exist. I worked for the Flight Operations Directorate (FOD) back then, which later split into the Flight Crew Operations Directorate (FCOD) and MOD. The world did not change when these organizational changes took place. These changes were merely necessary to better align us with our mission (which was, at the time, to fly the Shuttle forever and go on to do other, bigger things as well).

While it is a great and necessary thing to build comradery and esprit de corps around an organization like MOD, I believe that it is more important to rally the troops around a mission. So long as the organization supports the execution of that mission—or vision—then the synergy is ideal. As soon as the leadership finds that they have to change the mission in order to make it better align with the organization, then they have a problem. It needs to be the other way around. All organizations should be molded to fit the mission, and not vice versa.

People used to work at Johnson Space Center (JSC), putting in very long days and working weekends, because they believed in the mission— whether it was flying the first men in space, going to the moon, or flying the first reusable spacecraft. Even building the ISS was a great mission that had lots of dedicated people putting in excessive amounts of personal time. But later, we spent an inordinate amount of manpower trying to come up with ways to preserve our organization, grow our organization, or organize our organization—with few mission results to show for it. Lean and mean—that is how you accomplish the missions.

It should be noted that just a few years after I left the agency, JSC did, in fact, change up the organizational structure of operations again, this time recombining the Flight Crew Operations Directorate with the Mission Operations Directorate. In essence, they came back full circle to where they were when I joined JSC in 1980. The new Flight Operations Directorate has been working as a single entity now for a number of years, just like we did in the early days of the Shuttle program. This doesn't mean that MOD was a bad idea; it was a good idea for its time. But times change, and organizations need to change along with those times—and the missions.

Process vs. Product

About the time I started to get into a management role at NASA (as a lowly section head), the agency started looking for the latest management tools, theories, and practices. In my early days, it was clear that everyone in the organization was focused on technical matters—getting the job done, making sure our technical products and people were up to the task, and getting missions flown. Staff meetings were about those technical issues, vehicle status updates, training schedules, and so forth. Every bit of information we handled was mission related. Then things changed. We know why: the Shuttle was going to be an ongoing operational program "forever," and we felt that we needed to change from an R & D model to a production model for the human spaceflight operations organization. So we started looking at how business ran factories and production organizations. And this meant we started to focus on the processes more than we focused on the product.

Staff meetings became purely about process development. The theory is simple: if you fully understand the process, and document it well, you can create a "black box" into which you dump raw material, and out the other end will come finished, quality products. A product can be a fully trained flight control team. It can be a mission. It can be a new simulator. As far as management is concerned, the job becomes one of feeding the appropriate stock into the black box and making sure the box keeps running. Those managing the box no longer need to know or care about what is going on inside the box. It is far easier to manage than to create—and creation is what is going on inside the box. Humans tend to want to take the easy path.

The problem with this, of course, is that the easy path rarely, if ever, gets you anywhere. The truly creative people who advance an organization or field are, unfortunately, limited in number. We like to think that everyone we hired had the capability to be a star, but sadly, not everyone does. It is a subset of the employee population who can come up with new ideas and lead the way. The rest "feed the box." But when the focus ends up on the box itself, rather than on what actually goes on *inside* the box, the creative people are left out of the day-to-day operation. As a result nothing new gets done. Nothing new gets created. And the organization stagnates. If the box doesn't work, the organization no longer has the people who know enough about how it works to fix it or make it better. Attempts at process improvement fail, with thousands of man-hours spent for nothing—because it takes creative people to make it better, and those creative people need to be focused on the product, not the process.

I saw this almost every day in the last decade (or more) of my NASA career in our operational training organization. We spent a huge amount of time in defining the training process, and that time was being spent by our senior people who could just as well have been using their time and technical expertise to actually train folks. The overhead associated with process development was enormous, and the resulting training products just weren't that good. In the end, we were not in a mass production business; we were a boutique shop, and we didn't exist to produce millions of computers or cell phones, or cars. Our business was to make custom units, each one different, each one needing to be adaptable to changing situations. We needed to look at organizational models that emphasized this and then use our experienced people to actually conduct the technical training—not just build boxes that don't seem to be very effective. As I left the agency, these needs were being examined—a sign that organizations can at least be introspective enough to realize that they might need to be fixed.

The Need for Mentors

With very few exceptions, flight controllers are made, not born. In order for an engineer to become a flight controller, they need some form of instruction. In particular they need active participation in their development by senior, accomplished instructors who have "been there and done that." In the aviation world, the good instructors are older, seasoned, experienced pilots who pass on not just the physical skills of flying but also the lessons of many thousands of hours that help to build good judgment. At the highest levels of airline training, instructors are frequently pilots who have popped off the top of the stack due to the mandatory retirement age or a medical issue. These experts continue to earn their living in aviation by providing training for the next generation. It's a progression that sits in contrast with MOD's interesting history of staffing the training division with the newest, least experienced, fresh-out-of-school folks who—while full of good intent and energy—have no idea how to teach flight control skills because they had never been a flight controller.

To be fair, the training organization grew out of a need to train astronauts, not flight controllers. In the earliest days of human spaceflight, the crew was involved in defining their own training needs. The engineers followed along, supporting those needs with training devices, simulators, and classes. Flight controllers took care of training themselves since they were the ones developing the equipment and techniques to fly the missions. As programs, systems, and operations became more complex, the training organization grew such that it began providing integrated training for everyone involved in the operation. But it never really absorbed flight controllers who could teach the necessary basic skills needed by young, developing controllers. So we ended up with the excited new folks who had come to learn the systems and transfer that knowledge to flight crewmembers, but who were out of their depth when trying to teach the subtleties of flight control to engineers who had more experience than they had.

Over the years, MOD codified the values and standards that we expected flight controllers to meet, but we were less than successful at actively teaching these to every candidate who came along. Only by giving clear examples to follow, and actively helping new students practice and iterate on these skills, could we eventually produce the quality of individual that we expected to be sitting on console. This process takes time, and it is hard to make the time shorter because console expertise is not easily quantifiable. Using experienced instructors (which means senior flight controllers or Flight Directors in an active, instructional role) to conduct this training is necessary but expensive. The right people are few and it is hard to spread them around. As a result, there was the high number of failed certification attempts for FCR flight controllers. Although students had gone through the prescribed number of training exercises, something was missing in what they actually learned, and they failed to meet the standards set by the ultimate authority—the Flight Director Office. Time and again, failures were not associated with a lack of systems knowledge, but rather with communication and teamwork skills—things that must be learned with experience.

One of the ways in which this experience had been obtained (in the past) was through Multipurpose Support Room (MPSR) experience. Not only did this give the FCR operator a chance to actively teach the skills needed for the advancement of the trainee while they were in the back room, it also gave the observant trainee a chance to see those skills being exercised (successfully or poorly) by other flight controllers. It is not uncommon to find that more can be learned by watching than by doing, and while this is not always true, there is much to be gained by watching another person succeed—or fail. If nothing else, seeing a poor performance and the resulting critique helps people make more of an effort not to find themselves in the same situation. This MPSR experience was never explicitly identified as a significant contributor to the later success of a flight controller; rather, MPSRs were seen as a necessary evil in order to provide support in a discipline with too much going on for one person to handle.

Unfortunately, one of the results of making flight control operations more efficient was the thinning down of flight control teams. For some disciplines this meant the elimination of the MPSR altogether. The front room flight controller was all there was. This meant that a new flight controller had to start out totally exposed to the big organization. Unable to shelter behind their FCR operator for the early part of their career, they did not have the opportunity to observe how things worked from a position of relative anonymity and safety. The mistakes they made were in the public forum. One way the organization tried to handle this was to create a sub-tier of integrated training, held off-line in miniature control rooms, using training organization instructors to play the roles of crew-members as well as instructors. Unfortunately, this still didn't give the new flight controllers a real playground in which to mature their skills because the experienced operators were too busy flying the actual mission to spend much time in these sessions. So when the young flight controllers were exposed to the world of real integrated sims, they suffered poor pass rates on evaluations.

Training costs time, and that means money. But without proper training you end up with an organization of people who don't have the tools and experience to do the job. Money and time spent up front, early on in a person's career, pays dividends later. It is shortsighted for any organization to forgo that up-front investment.

Expert Leadership

One thing that was almost universally true in the old days of Flight Operations (which begat Mission Operations) was that the Section Heads and Branch Chiefs (and Division Chiefs for that matter) had all been experts in the type of work that their organizations performed. Group Leads in the Systems Division had been FCR systems flight controllers. Group Leads in Flight Dynamics had been Flight Dynamics Officers. Section Heads in the training organization had been senior instructors or Sim Sups. Now it is true that folks moved between disciplines, so that the Data Processing Systems Section Head might have been a Propulsion Systems Officer (although more often than not, the promotion was within a discipline). Nevertheless, the Section Heads were all expert flight controllers, and they could accurately judge the skill levels of their workers because they had been in those trenches themselves.

In the latter-day MOD, we had numerous people in leadership positions who had never sat on console—or at least who never did it successfully at a high level. We sometimes had people passing judgment on certification issues when the skills they were examining had never been examined in themselves. I believe that this was one of the primary reasons why we had a dismal "pass" rate on certifications for some FCR positions later in the program—people were not well prepared for their cert runs because the people responsible for getting them ready didn't really have the skills themselves. This was not universally true, of course—we had many group leads and senior managers who were expert flight controllers, and it showed. But we had a lot of less-experienced managers, and their results showed it as well.

The Marines have a tradition that everyone starts out as a rifleman. In a war situation, every Marine is therefore *still* a rifleman—from the private up to the general in the theater. Everyone is expected to fight if the need arises. I believe that Mission Control was a stronger organization with a better reputation when we were closer to that model. I believe that we needed to try harder to fill the leadership positions with people who had the experience that we expected from their followers. MOD drifted too far from the ideal and needed to correct its course back toward those who had proven that they understood what went on in real-time operations. Workers will always have a greater respect for management when an organization does that. Consequently, morale—and dedication to the mission—will be measurably higher.

The Role of Engineering in Real-Time Operations

Manned Space Flight Operations began with a small team of folks working out of the Langley Research Center. This team was responsible for the creation of the idea of manned spaceflight in the United States. It all started with folks who wanted to send people in space. They needed hardware to do it and they were able to outline what they needed—but then those hardware requirements were handed over to the design engineers to build. (At least that is the ops-side view of things. Like physicists arguing with mathematicians over who is actually supporting whom, the engineering side will say that they dreamed up the concept and simply "hired" the ops guys to fly them. Neither side is wrong.)

The operations team really began by coming up with a flight plan for the Mercury missions—what it was that they were going to do—from the first ballistic lobs through to the all-day orbital missions. In order to support the missions from the ground, they had to create the tracking network that allowed the maximum amount of command and telemetry. (Dr. Christopher Kraft, the first NASA Flight Director, once told us that the initial Mercury tracking stations were situated where they were so that the ground could talk to the crew about every fifteen minutes because, in the days before radar coverage, that was how often pilots of that era had to check in with air traffic control.) They also needed to create the tools and ground systems to look at the data, generate the commands, communicate between stations, and compute trajectories. This all turned into the organization called Mission Control.

The Engineering Directorate and the contractors who built the hardware always provided the engineering support, but these folks lived in a different world from the Operations team. The Operations team constantly trained with the clocks running—they learned to operate at the speed of flight. They couldn't stand down to consider a problem; they needed the best answer they could generate in a short period of time. Engineering was (rightfully) used to taking the time necessary to do a complete analysis, a luxury that Operations didn't have. So Operations developed a culture of knowing as much as they could while extrapolating beyond the "known" to make the most educated decisions possible. This is an entirely different technical discipline than design engineering. The Mission Evaluation Room (MER) function is, obviously, vitally important to the real-time team, but it needs to be remembered that the real-time team was created for a reason.

There is nothing bad whatsoever about the engineering organization's approach to what they do; it is simply different than what is required in operations. In a similar way, the launch control team at the Kennedy Space Center had a different job. While they did have to deal with quick and precise timing, they could almost always go into a hold if things weren't exactly right. The vehicle was still bolted firmly to the earth, after all. Their job was to safe the situation, then talk about what to do next. It's not better, or worse—it's just different . . . and you need different cultures to do different things.

"Badge on the Table" Moments

With great authority comes great responsibility, and the Flight Director Office expected its members to take that responsibility seriously. From the very first day, we have to be aware that at some point in our leadership career we might very well be faced with a "badge on the table" moment—one of those cases where you are asked (or told) to do something that is contrary to your core beliefs regarding crew safety and/or mission success. It is imperative, in those cases, that a Flight Director be willing to simply say, "No—I won't do that!," knowing full well that it can or will cost them their position or their job. Everyone who takes on the title of Flight Director has faced, or will face, moments when they must stand up for important ideals, regardless of the personal consequences.

It is also important to know that I have only seen a few cases where a senior leader has been forced to step aside because they refused to do something that was against their conscience and technical beliefs or understandings. But it has happened. We demand honesty and integrity of our leaders, especially those who are given authority over important projects and people's lives. And we would be disappointed if those people didn't live up to the ideals that they set for themselves. Bad things happen when you break the trust of others in a high-risk environment, and frankly, we can't get rid of people like that quickly enough. But it is sad when a man or a woman of high principles has to take the path of resignation in order to make a point.

In the aviation world, pilots who fly for a living need to have the courage to stand up to their bosses and managers and be willing to say, "No sir, I won't fly this airplane today." Whether it is a maintenance issue, an unsafe load, or bad weather—it does you no good to keep a job if you're going to be dead because you did something unsafe. The same thing is true of those involved in Space Flight Operations at the highest level. If you want to be given the authority and responsibility of a Flight Director, then you have to be willing to publicly walk away from it if the situation demands it.

Spaceflight Is Hard

When it comes right down to it, although humankind has made it to the moon and has spent close to sixty years with the capability of flying in low-Earth orbit, spaceflight is still hard—and for the foreseeable future it will remain that way. The energy it takes to blast someone off the planet and return them to Earth is so large that the margin for error is razor thin. Until we can create an entirely new method of getting people into orbit, we must face that fact. Creative, critical thinking and robust design will be the driving force for our ongoing forays into this environment. Good leadership will be essential too, because without inspiration who will put forth the effort to take the risks?

The Space Shuttle flight program lasted slightly over three decades— longer than the average human generation. There were many flight controllers at the end who weren't born when STS-1 left the pad for the program's first flight. They had never known a time when humans couldn't get on a Space Shuttle and take it to space. Then the program ended. It ended not because the vehicles were no longer capable of doing the job, not because we couldn't really afford it, but because we, as a nation, lost the vision, the leadership that had made the program work for so long.

Yes, there were immense risks. Over the span of 135 missions we lost two Shuttles and their crews. These are very poor odds when you compare them to the safety records that we expert of modern transportation systems. But space transportation today can't be compared to the safety records of airliners, or ships, or any other type of terrestrial transportation. The energy involved in going to space is a couple of orders of magnitude greater, and therefore the risks are much harder to contain. Besides, those safer modes of transportation have millions—or billions—of miles behind them. Many lessons had been learned along the way, leading them to be incrementally safer all the time.

Early aviation had a dismal safety record. Most pilots died young, after only a year or two of flying. Fast forward to the early airliners—they had just as many problems, with crafts going down in bad weather or over dangerous terrain. Many lives were lost, but we didn't ground all the aircraft and turn our backs on flying. Instead, we recognized that if we made air travel reliable and relatively safe, it would add immeasurably to the human experience. Indeed, aviation has shrunk the globe to the point where we think nothing of shipping a few dollars' worth of silicon from one side of the planet to another so that we can watch cats doing funny things on little screens.

Space travel can be as revolutionary as airline travel—we simply don't know what we will find to be its real potential until we play with it awhile. Aviation was a novelty until men and woman did the hard work to make it useful. And now we can't dream of being without it. Yet we walked away from thirty years of flying winged, reusable spacecraft because the national sense was that it was simply too dangerous, that there wasn't enough benefit for the risk. That assessment was so very wrong. The benefits of space exploration are seen in every hour of every day by anyone who lives in a technological world. Microelectronics, remote communications and sensing, our knowledge of the planet that allows us to make fuller and better use of our resources—these all come from the space program.

Yes, there are multiple ways to fly into space. We started out sealing humans into the nose cones of ballistic missiles. Then we added ways for the humans to control those nose cones and maneuver in ways that allowed them to accomplish interesting and useful missions. With the Space Shuttle, we developed the ability to put wings into space to allow us to bring back a huge vehicle—and huge payloads—on a runway. Then we turned that vehicle around and used it again—over and over and over. The Shuttle was a magnificent machine, and it taught us a great deal about how to routinely access space. And then we retired it in favor of small capsules that are little more than space taxis for humans.

The real shame of retiring the Shuttle is not that we moved on to something else; it is that we moved on to something *less*. The technology of the new commercial crew vehicles is far better than what we had in Apollo, and the safety of ballistic reentry is solid. But the capability of returning large payloads—be they scientific experiments or products made in space—has been lost. Humans have always explored new lands not just for the wonder, but for the new resources. Cargo ships laden with exotic trade goods spread civilization across the planet. You couldn't have done the same thing with row boats.

The day that we stopped flying the Shuttle was the day that the knowledge base for flying winged, reusable spacecraft into, and back from, space began to erode. The people who knew how to make that happen drifted off to other endeavors—many of those outside the space program. Sure, we still put people into space using expendable launch vehicles and small crew compartments that come back on parachutes, and that is far better than not having access to space at all. But it is far from the space freighter that we had in the Shuttle, and a far cry from what any fan of predictive science fiction has looked forward to throughout their life.

I am a fan of the commercialization of space transportation. When I get on an airliner, I am fully aware that it doesn't belong to the government. I know it belongs to a private, for-profit corporation—a commercial entity. And I believe that we need to head that same direction in space travel. In order for commercial companies to get going, they seem to have to take the same learning steps that NASA did: starting out with capsules and hopefully moving onward to fully reusable spacecraft that can carry cargo both into and out of space. Commercialization is a good thing, and if that is where the leaders go with their visions, then we will follow—and ride their ships off the planet.

I am also a believer that exploration is a good way for humanity to cooperate as we push back the boundary of the unknown. The established way of doing this has always been through governments. I believe that space exploration is too risky and too expensive for a commercial entity to take on alone—not that I wouldn't welcome the wealthiest people in the word putting their money in the game and go to Mars. But wealthy people tend to get that way because they are fiscally smart, and the chances are that the first expeditions to the planets will be money-losing propositions. Or better put, these expeditions will be investments that may not pay dividends until many years in the future.

But I do believe that those dividends will come—eventually. If humankind wants to progress, we have to move into space. We probably will not be forced to leave Earth in any sort of time frame we can imagine, but we need to expand our realm to other planets that can give us more resources and the capability to live off of the home world. This realistically won't happen in my lifetime—I know that now. When I was young, I think I expected that we'd be on Mars by now. But the reality of global and national politics, coupled with changing desires and wills, has made that goal more distant. I am not even sure what language will be spoken by the first humans who step on to the Martian surface—there is no reason to believe that it will necessarily be English.

But I do believe we will eventually hear a human voice from the surface of Mars, and again from the surface of the moon, and hopefully from other planets as well. These voices may come from people who are funded by great government efforts, or they may come from people who are funded by commercial entities. Regardless, I personally hope that they will be there for the good of all humankind, because no matter who funds the project, just getting there will be the big triumph. I believe that because I believe it is in our nature to explore, to grow, and to push the boundaries of what we know. If we stop doing that, we will stagnate. We need to push ourselves out of this little valley in which we live, climb to the ridgeline, and look out across a broader expanse at challenges we have yet to see. There will always be frontiers—we simply need the will to go out and find them.

Index